THEORETICAL AND PHYSICAL PRINCIPLES OF ORGANIC REACTIVITY

THEORETICAL AND PHYSICAL PRINCIPLES OF ORGANIC REACTIVITY

ADDY PROSS
Department of Chemistry
Ben-Gurion University of the Negev
Beer-Sheva, Israel

A Wiley-Interscience Publication

JOHN WILEY & SONS, INC.

New York • Chichester • Brisbane • Toronto • Singapore

Library of Congress Cataloging in Publication Data:

Pross, Addy, 1945–
    Theoretical and physical principles of organic reactivity / Addy Pross.
      p. cm.
    "A Wiley-Interscience publication."
    Includes bibliographical references (p.   –   ) and index.
    ISBN 0-471-55599-1 (alk. paper)
    1. Chemistry, Physical organic. I. Title.
QD476.P755 1995
547.1—dc 20
                                          95-21389
                                             CIP

Printed in the United States of America

10 9 8 7 6 5 4 3 2 1

CONTENTS

PREFACE

Unlike many other branches of science, chemistry is a scientific continent. The areas of study that are embraced by the general term "chemistry" are unusually diverse, ranging from the highly theoretical and physical, through the traditional organic and inorganic disciplines, into material science and extending into biology. The enormous scope of chemistry has necessarily led to the situation that different parts of the subject have developed independently, so that specialists in one area of the subject are often quite unaware of developments in some other area.

Organic chemistry has also reflected this tendency toward compartmentalization. Whereas the fact that the underlying quantum mechanical basis for chemical behavior has been well understood by physicists and theoretical chemists for over one half a century, much of the progress in organic chemistry had been developed almost independently of that fundamental quantum base. In fact, physical organic chemistry, that field of study that focuses on problems of organic structure and reactivity, has historically utilized an empirical rather than a theoretical approach.

Some 30 years ago Woodward and Hoffmann, and Fukui, ushered in a new era in organic chemistry with their publications on frontier molecular orbitals (FMOs) and the importance of orbital symmetry. The significance of this work was that it dramatized that basic theoretical concepts derived from quantum mechanics are necessary for a proper understanding of experimental organic chemistry—that the phase of an orbital may determine whether a reaction in the laboratory will succeed or fail, and whether a molecule is likely to be stable or unstable. It was these chemists, in particular, who forged the modern link between theoretical and experimental chemistry.

As a young lecturer at Ben-Gurion University of the Negev in the early

1970s, I found myself swept along by the MO wave that followed the Wood-ward–Hoffmann–Fukui contributions. But in 1979 my interests were diverted to valence bond thinking when Sason Shaik joined the Chemistry Department at Ben-Gurion University. It was Sason Shaik who introduced me to electronic configurations and valence bond (VB) theory. He showed me that VB theory was not an out-dated theoretical approach that had been superseded by MO theory, but an alternative theoretical procedure whose potential remained largely untapped, and that a combined MO–VB approach to chemical problems was a particularly powerful one. And so the 1980s found Sason Shaik and myself collaborating on ways this potential could be utilized for an improved under-standing of organic reactivity and organic reaction mechanism. In this period we produced, jointly and separately, many papers that described and applied a model we termed the valence bond configuration-mixing (VBCM) model (when applied using configurations), or the state correlation diagram (SCD) model (when applied using states). More recently, we use a general term to cover both approaches: the curve-crossing model.

For me this collaborative adventure was very exciting. Having worked in the area of organic reactivity for some 15 years prior to this period, it was with amazement that I began to realize that I had been working in that area without having understood why most chemical reactions have a barrier! More remark-able still was the realization that the theoretical ideas required to understand barrier formation had been largely understood by the mid-1930s, but that these basic concepts had somehow been put aside by experimentalists as belonging to ''theoretical chemistry.''

So my intent in this book is to present a more unified approach to the problem of organic reactivity, which has developed from this VB period of study, and to bring about a merging of areas that are often treated as separate. Accordingly, traditional concepts in physical organic chemistry are presented together with the theoretical insights afforded by qualitative MO and VB theories. In addition, the reader will find that the text presents a detailed description of electron-transfer theory: an area traditionally associated with transition metal chemistry. However, as is described in some detail, electron-transfer reactivity and organic reactivity share many common features. In effect, the electron-transfer process is at the heart of many organic reactions, so that the Marcus and Taube theories of electron transfer, taken from physical and inorganic chemistry, become important elements in a framework for understanding organic reactivity.

The book is divided into three parts. Part A presents a description of MO and VB theories. The emphasis is on the qualitative aspects that can be applied to practical problems in organic structure and reactivity, leading up to a de-scription of why chemical reactions have a barrier, and the factors that govern the height of this barrier. Part B presents the key principles of physical organic chemistry and includes a mainly qualitative description of the Marcus theory of electron transfer. Finally, Part C presents an overview of the basic reactions of organic chemistry—nucleophilic and electrophilic substitution, radical and pericyclic reactivity—utilizing the theoretical framework presented in Parts A and B. The coverage in this section is not intended to be comprehensive, but rather aims to focus on recent developments in these areas.

The text is primarily intended for advanced undergraduate and graduate students as well as researchers in organic chemistry. It presumes a knowledge of basic organic mechanism, but assumes a very limited theoretical background. For this reason, the book presents, where possible, important theoretical concepts in a nonmathematical manner. For example, much of the theoretical discussion on electronic wave functions presents these functions in a chemical form—as resonance-type structures, rather than using the more abstract mathematical formalism.

I owe a debt of gratitude to many people. In particular, I must acknowledge Professor Lennart Eberson, Dr. Mikhail Glukhovtsev, Professor Henning Lund, Professor Robert Moss, Professor John Pinhey, and especially Professor Leo Radom and Professor Sason Shaik, for their many comments and criticisms of an early draft; the present version has benefited greatly from their input. I am also grateful to the Australian Research Council for the award of a Senior Research Fellowship, during whose tenure at the University of Sydney much of this book was written. Last, but not least, I wish to express my gratitude to Nella, my wife, for her encouragement and support, and especially for her dedicated art work that converted rough and often incomprehensible sketches into the computer figures used in the text.

ADDY PROSS

School of Chemistry, University of Sydney, Sydney, Australia and
Department of Chemistry, Ben-Gurion University of the Negev,
Beer-Sheva, Israel

PHYSICAL CONSTANTS AND CONVERSION FACTORS

Boltzmann's constant	**k**	1.381×10^{-23} J K^{-1}
Planck's constant	h	6.626×10^{-34} J s
Electronic charge	**e**	-1.602×10^{-19} C
		-4.803×10^{-10} esu
Avagadro's number	N	6.023×10^{23}
Universal gas constant	R	1.987 cal mol^{-1} K^{-1}
		8.314 J mol^{-1} K^{-1}

1 eV $= 1.602 \times 10^{-19}$ J $= 1.602 \times 10^{-12}$ ergs
1 cal $= 4.184$ J
1 eV mol^{-1} = 23.06 kcal mol^{-1} = 96.49 kJ mol^{-1}

PART A

THEORETICAL PRINCIPLES

1

FUNDAMENTAL CONCEPTS

1.1 INTRODUCTION

Since this book deals with the nature of chemical reactivity, it may be appropriate to begin our discussion with a few words that clarify what is actually meant by the term "a chemical reaction." A chemical reaction involves the reorganization of the nuclei and electrons that together constitute the particular chemical system. In the reactants the atoms are held together by the so-called valence electrons in a given arrangement that defines the molecular structure of the reactants. In the products, the very same atoms are held together, generally by the same valence electrons, but in a different geometric arrangement. So both the electronic "glue" and the relative positions of the atoms they hold together have undergone major reorganization. In normal chemical terminology this structural and electronic reorganization is described as the *breaking of old* and the *making of new chemical bonds*. One of the principal features of this book is the focus on physical rules that govern the way this general process occurs; in particular, how much energy is needed to overcome the activation barrier that almost invariably separates reactants from products. Ultimately, almost the entire question of chemical reactivity hinges on this simple question.

Since any chemical reaction involves the making and breaking of bonds, a detailed description of what a chemical bond is, needs to be considered. The theoretical description of structure and bonding in chemistry has historically been divided between two major approaches: *valence bond (VB)* theory and *molecular orbital (MO)* theory. The VB theory was first formulated by Heitler and London[1] in 1927, to describe the electron pair bond in the hydrogen molecule, and was then developed, primarily by Pauling,[2] as a general theory of chemical bonding. At the same time MO theory was being formulated, primarily by Hund,[3] Mulliken,[4] and Hückel.[5] The relative influence of these two theoretical approaches has changed enormously over the years. The VB theory and its simplified, pictorial extension (resonance theory) dominated chemical thinking in the 1940s and 1950s, while the MO theory grew enormously fashionable in the 1960s and 1970s. The changing status of the two theories over time is readily apparent from reading textbooks of the relevant periods. Thus Pauling and Wilson,[6] in their classic *Introduction to Quantum Mechanics*, deemed MO theory worth no more consideration than one and a half pages, while the rejection of VB theory during the 1960s and 1970s is exemplified by a popular text on chemical bonding published at the time, that almost apologetically introduces a brief discussion of VB theory in one of the last chapters in the book.[7] A more balanced viewpoint would be that MO and VB theory, rather than representing two alternative, and sometimes conflicting, approaches to chemical structure and reactivity, are actually complementary, and that each theory is particularly suited to the consideration of certain aspects of chemical behavior. In fact, in the course of this book, it will be shown that VB theory, because of the emphasis it places on electron bond pairs, has some advantages over MO theory in its ability to qualitatively deal with problems of chemical reactivity. The advantages of MO theory become evident when problems of stereochemistry and symmetry are considered, and in computational chemistry. The differences between MO and VB approaches, and their particular capabilities, will be discussed at appropriate points in this book.

1.2   THE SCHRÖDINGER EQUATION

The behavior of atoms and molecules, and hence the basis of chemical theory, is governed by quantum mechanics. Quantum mechanics postulates that different properties of a system, termed *observables*, may be found by solving equations of the general form

$$G\Psi = \gamma\Psi \qquad (1.1)$$

where γ is the particular observable that is being calculated, Ψ is a mathematical expression termed the *state function*, which defines the system in a given state, and G is a mathematical *operator*, whose nature depends on the observable being calculated. Since the mathematical form of the state function Ψ resembles the wave equations of classical physics, these functions are commonly termed *wave functions*. In other words, Eq. (1.1) states that when a certain mathematical operation is carried out on the state function, the result is a product of the state function and the observable itself. In the language of quantum mechanics Ψ is termed an *eigenfunction*, while γ is termed an *eigenvalue*. Thus a set of eigenfunctions is obtained by solving equations of this type (Eq. 1.1), and from them, the theoretical values of the observables of the system—one value for each eigenfunction.

Of particular importance to chemists is the *energy E* of a chemical system, so when energy is the observable that is calculated in Eq. (1.1), the equation is termed the Schrödinger wave equation.

$$\mathcal{H}\Psi = E\Psi \tag{1.2}$$

The operator $\mathcal{H}$ (whose precise nature we need not consider) is an energy operator, termed the Hamiltonian, that considers all of the kinetic and potential energy contributions to the electronic energy. So, in principle, solution of the Schrödinger equation for a given system leads to a set of electronic wave functions Ψ_i, of energy E_i which represents the set of allowed electronic states (ground and excited) for the system. Despite the superficial simplicity of Eq. (1.2), in practice solution of the Schrödinger equation is mathematically awesome. This stems from the complicated mathematical form of $\mathcal{H}$ and Ψ. In fact, the *exact* solution to the Schrödinger equation can only be obtained for one-electron systems, such as the H atom or the H_2^+ molecular ion. For this reason approximate methods, which make the solution of the equation more tractable, must be followed. However, even these approximate methods can be mathematically complex and may still require enormous amounts of computer time. Dirac's famous aphorism[8] states the problem explicitly:

> The underlying physical laws necessary for the mathematical theory of a large part of physics and the whole of chemistry are thus completely known, and the difficulty is only that the exact application of these laws leads to equations much too complicated to be soluble.

And here lies a major paradox of chemistry: On the one hand, chemistry is ultimately no more than an exercise in applied mathematics. On the other hand, the complexity of the mathematical process, to a large extent detaches the procedure from human understanding. For this reason chemistry is awash with a wide range of qualitative and semiquantitative theories and concepts. The concepts of the chemical bond, substituent effects, orbitals, hybridization, electronegativity, functional groups, and so on, are obvious examples. If one were to take Dirac's aphorism literally, then at least in principle, quantum mechanics would make many of these ideas superfluous. Yet, despite the obvious limitations of these concepts, it is only through their application that a chemist achieves "understanding," imperfect as this might be, and is the basis by which he/she performs his/her art.

1.3 ATOMIC ORBITALS

Just as atoms represent the physical building blocks from which molecules are constructed, so atomic orbitals (or simply, AOs) are the conceptual building blocks from which all the chemists' concepts of structure and reactivity derive.

When the Schrödinger equation is applied to a single atom, its solution yields the mathematical functions ϕ which define the set of AOs. The best known AOs are the atomic s, p, and d orbitals illustrated in **1.1**. Shaded areas indicate regions of the orbital where the value of the wave function is positive, while unshaded areas indicate regions in space where the value is negative. Planes that define coordinates in space where the wave function has zero value and separate positive regions from negative regions, are termed *nodes*. Thus the simplest s, p, and d AOs possess 0, 1, and 2 nodes, respectively. As we will see, it is the total set of AOs that forms the input for both MO and VB theories.

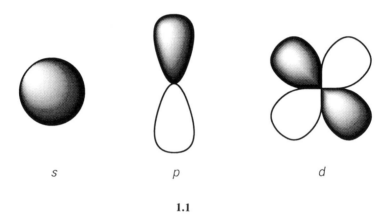

s   *p*   *d*

1.1

1.4   THE COVALENT BOND IN THE HYDROGEN MOLECULE

As noted in the introduction, a chemical reaction involves the breaking and making of bonds. For this reason a proper understanding of the chemical bond must underlie our attempt to understand the factors governing structure and reactivity. Since the most ubiquitous of all bonds is the covalent bond, let us begin with a brief MO and VB description of the covalent bond in the simplest of all molecules: the hydrogen molecule.

1.4.1   MO Description

In a hydrogen atom in its ground state, the single electron resides in a spherical *1s* orbital. When two hydrogen atoms, labeled A and B, respectively, approach one another, each electron now feels not only the nuclear potential associated with its own nucleus, but that of the second nucleus as well. In such a situation the spherical electronic distribution of the *1s* orbital is no longer appropriate. Now each of the two electrons will distribute themselves around the two nuclei in a way that is governed by the *combined* potential generated by the two hydrogen nuclei. The wave function that defines this new electronic distribution about the two nuclei is called a *molecular orbital* or MO, since it is associated with a molecule and not an atom. So the essence of the MO description may be stated as follows: *Electrons within a molecule occupy MOs rather than AOs; these MOs reflect the combined electric fields of all the atoms within the molecular system.*

How can the wave function for this new orbital be obtained? According to the procedure termed the *linear combination of atomic orbitals* (LCAO), which will be described in detail in Section 2.1.1, a possible MO wave function may be obtained by simply taking a linear combination of the atomic orbital wave functions on Atoms A and B, ϕ_A and ϕ_B, respectively. Such a function is likely to reflect the combined influence of the two AOs. Since a linear combination may be obtained both by adding and subtracting the atomic wave functions, the LCAO procedure for two AOs yields *two* MOs, σ and σ^*, as expressed in Eqs. (1.3) and (1.4). The σ orbital comes about from the *in-phase* mixing of the two AOs, while the σ^* orbital comes about through the *out-of-phase* mixing of the two orbitals. (*N* is a normalization constant, whose purpose it is to ensure that the total electron density defined by the wave function is equal to 1; see Section 2.1.3.)

$$\sigma = N(\phi_A + \phi_B) \tag{1.3}$$

$$\sigma^* = N(\phi_A - \phi_B) \tag{1.4}$$

Most importantly, the σ orbital is *stabilized* relative to the two *1s* orbitals. Consequently, placing the two available electrons into this lower energy σ orbital leads to a reduction in electronic energy; that is, a covalent bond between the hydrogen atoms is formed. A detailed description of orbital mixing will be described both quantitatively and qualitatively in Chapter 2.

1.4.2 VB Description

The VB approach to the covalent bond is very different from the MO approach because it focuses on electron-pair bonds between atoms, even in complex molecules. Thus, according to the VB approach, *electrons are associated with AOs, not just in atoms, but in molecules as well.* Historically, the VB approach preceded the MO approach and indeed, Heitler and London's description of the covalent bond was a landmark event in chemical thinking because it provided the first physical understanding of the nature of the covalent bond. By applying the principles of the just-discovered quantum mechanics to the hydrogen molecule, Heitler and London were able to demonstrate theoretically that the wave function describing an electron pair shared between two atoms, leads to the formation of a stable bond between the two atoms. The complete Heitler–London wave function Ψ_{COV}, for the covalent bond in the ground-state hydrogen molecule, is given by

$$\Psi_{COV} = N[\phi_A(1)\phi_B(2) + \phi_B(1)\phi_A(2)][\alpha(1)\beta(2) - \beta(1)\alpha(2)] \tag{1.5}$$

where ϕ_A and ϕ_B are the *1s* AO wave functions for each of the two hydrogen atoms A and B, N is a normalization constant, α and β are the electron spin functions, and the integers 1 and 2 are labels for each of the two electrons.

This equation, rather imposing in its appearance, is actually quite straightforward; it merely transcribes a simple physical picture that specifies the location of the electrons and their spins into a mathematical formalism. The equation is constructed from two parts: a *spatial* part and a *spin* part. The spatial part, which appears within the first set of square brackets, contains two components $\phi_A(1)\phi_B(2)$ and $\phi_B(1)\phi_A(2)$, which specify where the two electrons are located: one in the AO associated with hydrogen atom A and the other in the AO associated with the hydrogen atom B. The reason there are two components in the spatial part of the wave functions is to allow for the fact that the two electrons are *indistinguishable*, and hence may be interchanged; that is, the wave function *must* allow for electrons 1 and 2 to reside on either atom A or atom B.

The second part of the wave function is the spin part. It specifies that the two electrons have opposite spins (one α, one β), and, in analogy to the spatial component, the spins are also interchangeable (i.e., the electron on atom A may be either of spin α or β). So the fact that Eq. (1.5) is made up of two spatial components and two spin components, rather than just one of each, is to allow for the interchange of any two electrons. *So all Eq. (1.5) states in mathematical terms is that there is one electron associated with each of the two H atoms, and that the two electrons have opposite spins.*[a]

Why does the VB wave function of Eq. (1.5) and the corresponding MO wave function [Eq. (1.3) with an occupancy of two electrons] describe the formation of a bond? Quite simply, because substituting these wave functions into the Schrödinger equation (1.2) leads to a *lower* energy when the two H atoms are in close proximity than when they are separated. To state that there is a chemical bond between two atoms (or two groups) is just to state that the energy of the system when the atoms are in close proximity is lower (for whatever reason) than when the two moieties are separated.

The Heitler–London wave function [Eq. (1.5)] is the basis of VB theory,

---

[a]There is actually an additional condition that is imposed on an electronic wave function—that it be *antisymmetric*. This means that interchanging any two electrons within the wave function must give the negative of the original wave function. This condition is imposed since states that are *symmetric* have never been observed spectroscopically. So while quantum mechanics would appear to give rise to symmetric as well as antisymmetric wave functions, it appears nature has decreed that only the antisymmetric functions correspond to proper physical states. The antisymmetry property may be expressed mathematically by

$$\Psi(1, 2) = -\Psi(2, 1)$$

and is the formal representation of a well-known chemical principle (the Pauli exclusion principle), which states that two electrons with the same spin *cannot* occupy the same orbital.

Exchanging electrons 1 and 2 in Eq. (1.5) generates the negative of the original function confirming it to be antisymmetric as required.

and may be used to describe *any* two-electron covalent bond. Its approach, in contrast to that of MO theory (where electrons are considered to be *delocalized* over the entire molecule), is to describe bonds as *localized*—the two electrons that make up the particular bond are associated with those two atoms that are bonded together. Of course, VB theory can also be used to describe the electronic structure of large molecules that contain many covalent bonds, and not just those with a single bond. However, in order to illustrate the way in which VB theory (and, as we will see, MO theory also) considers the electronic structure of a many-electron system, we first need to consider the concepts of *electronic configurations* and *electronic states*.

1.5 ELECTRONIC CONFIGURATIONS

An important concept in our description of a molecule or system of molecules is the *electronic configuration*. *When electrons within a molecule (or system of molecules) are allocated to specific orbitals (either AOs or MOs) we obtain an electronic configuration for that molecule (or system of molecules).* So the electronic configuration is a theoretical construct that tells us *where* in the molecule the electrons are located. An electronic configuration is described mathematically by a wave function, normally labeled by the symbol Ψ. Any electronic configuration may be described using MO or VB type wave functions, so we need to consider each in turn.

1.5.1 MO Configurations

As discussed above, the hydrogen molecule MOs are the σ and σ^* orbitals, obtained by the in-phase and out-of-phase linear combination of the hydrogen $1s$ AOs. To generate a set of *MO configurations* from this set of MOs, the two electrons associated with this molecule now need to be allocated to these two MOs. This may be done in several different ways, and each possibility constitutes a different MO configuration. Various possibilities are illustrated in Fig. 1.1. Thus the *ground configuration* Ψ_0 is obtained by placing both electrons into the σ orbital (Fig. 1.1a), while two possible *excited configurations*, Ψ_1 and Ψ_2, where one or both electrons are excited into the higher energy σ^* orbital, are depicted in Fig. 1.1b. The three configurations are depicted mathematically by

$$\Psi_0 = \sigma^2 \tag{1.6}$$

$$\Psi_1 = \sigma^1 \sigma^{*1} \tag{1.7}$$

$$\Psi_2 = \sigma^{*2} \tag{1.8}$$

The superscript index, that appears adjacent to the orbital designation, indicates the number of electrons residing in that orbital.

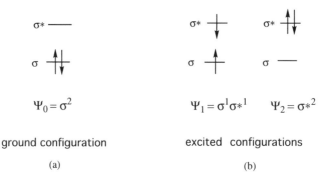

ground configuration   excited configurations

(a)   (b)

Fig. 1.1 (a) Ground MO configuration Ψ_0. (b) Excited MO configurations Ψ_1 and Ψ_2, for the H_2 molecule.

More generally, for any system, the set of MOs, ψ_1, ψ_2, ψ_3, $\cdots$ ψ_N (the way we generate a set of MOs is described in Chapter 2), may be schematically represented as a set of energy levels. This is illustrated in Fig. 1.2a for a set of six MOs. Lowest in energy is ψ_1, with the other MOs being ranked above it in increasing order of energy. Creation of different MO configurations is generated by allocating the electrons to the MO set in different ways. Each arrangement of electrons within the set of MOs generates a unique electronic configuration. If the electrons are placed in the available MOs in the *lowest* possible energetic arrangement, we obtain the *ground configuration* Ψ_0. This arrangement is illustrated in Fig. 1.2b for the arbitrary case of six electrons.

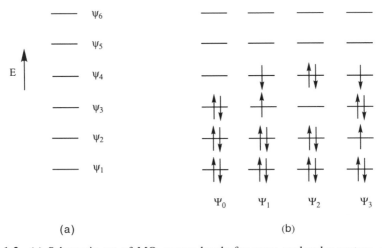

Fig. 1.2 (a) Schematic set of MO energy levels for some molecular system. (b) A selection of MO configurations $\Psi_0 - \Psi_3$ for a six-electron system based on the MO levels shown in (a).

The wave function that describes this ground configuration is just

$$\Psi_0 = (\psi_1)^2(\psi_2)^2(\psi_3)^2 \qquad (1.9)$$

while excited configurations of higher energy may be generated by promoting the electrons to higher energy MOs. Thus Ψ_1 is defined by

$$\Psi_1 = (\psi_1)^2(\psi_2)^2(\psi_3)^1(\psi_4)^1 \qquad (1.10)$$

and may be obtained from Ψ_0 by promoting an electron from ψ_3 to ψ_4. A selection of several electronic configurations, out of the entire set that may be generated in this way ($\Psi_0 - \Psi_3$), is illustrated in Fig. 1.2b.

1.5.2 VB Configurations

Just as we constructed a number of MO configurations for the H_2 molecule (see above), we can apply the VB approach to construct a number of VB configurations for that same molecule. The difference between the two approaches rests on the fact that in the VB approach the electrons are allocated to AOs, while in the MO approach the electrons are allocated to MOs. The Heitler–London wave function for the hydrogen molecule [Eq. (1.5)] is such a VB configuration, since one electron is allocated to the *1s* orbital of one H atom, and the other electron to the *1s* orbital of the other H atom. Two additional VB configurations, representing *ionic* forms are

$$\Psi_{ION1} = N[\phi_A(1)\phi_A(2)][\alpha(1)\beta(2) - \beta(1)\alpha(2)] \qquad (1.11)$$

$$\Psi_{ION2} = N[\phi_B(1)\phi_B(2)][\alpha(1)\beta(2) - \beta(1)\alpha(2)] \qquad (1.12)$$

In Ψ_{ION1} the two electrons reside in the *1s* orbital of one H atom, while in Ψ_{ION2} the two electrons reside in the *1s* orbital of the other H atom.

The mathematical representations of VB configurations, as depicted in Eqs. (1.5), (1.11), and (1.12), are chemically obscure, at least when viewed through the eyes of many experimental chemists. Since we will find these VB configurations to be of profound importance in our subsequent discussions, it becomes desirable to find a more *chemical* means of representing these configurations. So let us now describe how VB wave functions can be represented in pictorial fashion using resonance-type structures, an approach used extensively throughout this book.

1.5.3 Pictorial Representation of VB Configurations

As noted above, electronic configurations are mathematical expressions. Indeed it is the fact that theoretical and experimental chemistry are each based on such different languages (mathematics and structural formulas) that has kept these two intimately related fields from merging effectively. For this reason it becomes important to find a nonmathematical means of representing, what are in essence, mathematical formulations; we somehow need to assimilate basic theoretical concepts into the language of experimental chemistry. The approach we take here is to represent VB configuration wave functions in a pictorial form, using resonance-type structures. Our aim is to capture the mathematical essence of the wave function, while employing the language of experimental chemistry. Let us illustrate this pictorial representation for the three VB configurations of the H_2 molecule [Eqs. (1.5), (1.11), and (1.12)].

The Heitler–London wave function for the H_2 molecule, depicted in mathematical form in Eq. (1.5), is represented in structural language simply by placing the valence electrons adjacent to the appropriate atom. Accordingly, Eq. (1.5) becomes

$$\Psi_{COV} = (H \cdot \uparrow\downarrow \cdot H) \xleftrightarrow{(-)} (H \cdot \downarrow\uparrow \cdot H) \qquad (1.13)$$

The two terms in Eq. (1.5), which allow for spin exchange, appear as two structures in the pictorial representation, and are connected by a resonance arrow to signify the mixing of these two forms. The negative sign above the resonance arrow indicates that it is the *negative* combination of the two spin forms, $\alpha\beta$ and $\beta\alpha$, that represents the singlet ground configuration of the hydrogen molecule [as expressed in Eq. (1.5)]. To simplify matters even further we may reduce the structural representation of Eq. (1.13) to

$$\Psi_{COV} = H \cdot \ \cdot H \qquad (1.14)$$

In this simplified representation it is understood that the two electrons are of opposite spins and are coupled together in a bond pair. *The importance of this structural representation of the Heitler–London wave function cannot be overemphasized.* It will recur frequently throughout this book and is one of the cornerstones in all our subsequent discussions on reactivity. (An alternative representation of $H \cdot \ \cdot H$, which we will not employ here, but which emphasizes the fact that the two electrons are spin paired, is $H \cdot - \cdot H$.)

In similar fashion, the wave functions for the two ionic configurations of the H_2 molecule, expressed mathematically by Eqs. (1.11) and (1.12), may be represented pictorially as

$$\Psi_{ION1} = H:^- \ H^+ \qquad (1.15)$$

$$\Psi_{ION2} = H^+ \ :H^- \qquad (1.16)$$

The reader will see immediately that for a molecule such as H_2, the pictorial representations of VB configurations [Eqs. (1.14–1.16)], reduce to just the simple resonance structures of classical organic chemistry. However, the resonance formulation need not be restricted to describing the electronic structures of stable species (benzene, the enolate ion, and the allyl cation are famous examples). As we discuss in Chapter 3, the same VB methodology can be applied to a reacting system made up of *more* than a single molecule, and may be used to build up a reaction profile and to describe transition states.

This ability to represent VB configurations with resonance-type structures gives the VB methodology an important advantage over the MO method; VB configurations may be couched in the language of organic chemistry and therefore can be made to closely resemble the Lewis structures that, to this day, form the basis of the chemist's way of thinking about, and representing chemical structures. This point needs to be emphasized: Each time we draw a Lewis structure for a simple organic molecule, for example, **1.2** and **1.3** for CH_4, we are actually communicating in VB language. Given that the lines joining atoms in such structural formulas are understood to represent bond pairs, then the structural formula of **1.2** and the Lewis electron dot picture **1.3**, are both useful representations of the main VB configuration of the CH_4 molecule.

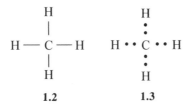

1.2	**1.3**

1.5.4 Electronic Configurations for Molecular Systems

Electronic configurations may be constructed, not just for single molecules, but for a *molecular system* (a system of more than one molecule). Consider a system composed of a donor molecule $N:^-$, and an acceptor molecule $R-X$. This system is characterized by four key electrons: the lone pair on the nucleophile and the two electrons of the $R-X$ bond. One can construct a set of electronic configurations for this molecular system by rearranging these four electrons among the different orbitals. This construction can be done using both MO and VB methodologies.

The MO configurations are derived by rearranging the four electrons among the three key MOs of this system: The so-called *frontier molecular orbitals (FMOs)*. The FMOs are just the *highest occupied molecular orbitals (HOMOs)* and *lowest unoccupied molecular orbitals (LUMOs)* of the system. In the case of $N:^- + R-X$, the FMOs are the lone-pair orbital n on the nucleophile, and the σ and σ^* orbitals of the $R-X$ molecule. A set of configurations (one ground and two monoexcited) is illustrated schematically in Fig. 1.3 and may be labeled with the aid of the D (donor) A (acceptor) terminology, where the nucleophile N^- is the donor and the $R-X$ molecule is the acceptor. Thus the three configurations shown in Fig. 1.3 would be designated as follows:

$$DA = n^2\sigma^2 \tag{1.17}$$

$$D^+A^- = n^1\sigma^2\sigma^{*1} \tag{1.18}$$

$$DA^* = n^2\sigma^1\sigma^{*1} \tag{1.19}$$

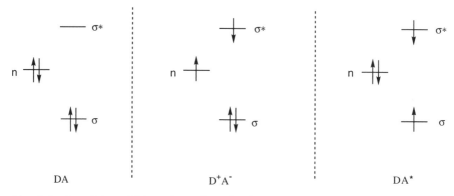

Fig. 1.3 Set of three MO configurations for the system, $N:^- + R{-}X$ obtained by occupying the frontier MOs with the four valence electrons of the system. Configurations are labeled using the D (donor) A (acceptor) terminology. Here DA is the ground configuration; D^+A^- and DA* are two singly excited configurations.

Here DA is the MO ground configuration, while D^+A^- is the MO charge-transfer configuration, obtained by the transfer of an electron from the donor HOMO into the acceptor σ^* orbital. The locally excited configuration DA* is obtained through the excitation of an $R{-}X$ σ electron into the σ^* orbital. Of course, additional MO configurations of higher energy can be constructed by the excitation of *two* electrons.

The set of VB configurations for this system may be generated by rearranging the four valence electrons among the AOs (actually hybrid orbitals) of the bonding atoms. The set of configurations is shown in Structures **1.4–1.7**, by way of the pictorial representation discussed previously. It is apparent that the VB configurations are more closely related to the language of structural chemistry than the MO configurations.

Unfortunately, none of the above configurations (MO or VB), provides an *accurate* electronic description of the system $N:^- + RX$ [though inspection of the set reveals immediately that **1.4** is a reasonably good electronic representation of this system; it displays the nucleophile lone pair as well as the Heitler–London $R{-}X$ bond pair.] The *actual* description of the electronic arrangement of a system is governed by its *electronic state.* Let us now consider this concept, and consider the relationship between configurations and states.

$$N:^- \ R\cdot \ \cdot X \qquad N\cdot \ \cdot R : X^- \qquad N:^- \ R^+ \ :X^- \qquad N\cdot \ R:^- \ \cdot X$$
$$\quad \textbf{1.4} \qquad\qquad \textbf{1.5} \qquad\qquad \textbf{1.6} \qquad\qquad \textbf{1.7}$$

1.6   ELECTRONIC STATES

The concept of the electronic state is most important, since ultimately, it is the electronic state that governs the electronic behavior of a system, and through it, much of its structural and reactivity characteristics. *An electronic state is just an arrangement of the electrons in the molecule (or system of molecules) that is allowed by the laws of quantum mechanics.* A molecule (or system of molecules) may exist in any one of a number of electronic states. These include the electronic *ground state*, as well as a number of possible *excited states.*

Mathematically each electronic state can be described by a wave function, termed a *many-electron state wave function*, or more briefly, a *state wave function*, and designated by the symbol Ψ (the same symbol used for the electronic configuration wave function). Thus the state wave function describes the total arrangement of all the electrons within the system, and from it one can, in principle, derive molecular properties, the most important ones being energy and electronic distribution, for the state itself. Since each state of a system is described by a different wave function, the wave functions for the different states may be labeled Ψ_g, Ψ_{e1}, Ψ_{e2}, Ψ_{e3}, and so on, where Ψ_g is the ground state, Ψ_{e1} the first excited state, and so on.

Chemical reactions may involve different electronic states of a reacting system. The most common reaction class are *thermal* or *ground-state reactions*, which involve the transformation of molecules in their *ground* state to product molecules, also in their ground state. *Photochemical reactions* involve the initial excitation of the ground-state molecules to some accessible excited state, which then subsequently react to give the products of the reaction, normally in the electronic ground state. The phenomenon of *chemiluminescence* involves the ground-state reaction of molecules to generate products in an excited electronic state, which can then drop down to the ground state with concomitant emission of radiation.

This book will deal primarily with the first and largest class—reactions in which *ground-state* reactants are transformed into *ground-state* products along a *ground-state* reaction surface. Since the state wave function is nothing other than the mathematical formulation of the physical description of the chemical system, any insight we can obtain into the nature of the state wave function will translate directly into a more basic understanding of both chemical structure and reactivity. So our first goal will be to define the nature of the ground-state wave function; this will provide us with insight into questions of structure. Our second goal will be to explore how the ground-state wave function *changes* during the course of the reaction from reactants through the transition state and into products. This treatment will provide us with insight into problems of reactivity.

As we have just noted, in order to be able to fully describe a particular state, we need to find the state wave function. Unfortunately to do this with total precision, that is, to find the *exact* state wave function, is impossible. The best we can do is to make increasingly more accurate representations of a particular state wave function. So how do we begin?

The simplest representation of a state wave function is some appropriate configuration wave function. An electronic configuration may serve as an approximate description (or label) for a particular electronic state, even though it is *not* an exact description of that state. In fact, whenever we think about a molecule we intuitively apply a configuration description to that molecule. As an example, the MO ground *configuration* wave function for the hydrogen molecule σ^2 is a useful MO representation of the ground *state* of the hydrogen molecule, even though it is not a precise representation of the state. In similar fashion the VB configuration, H· ·H [Eq. (1.14)] is *also* a useful (though approximate) VB representation of the ground state of hydrogen. In Section 1.7 we will see *why* an electronic configuration is only an approximate description of a state, and the means by which one can obtain increasingly more accurate state wave functions by the mixing of configuration wave functions.

1.7 GENERATING STATE WAVE FUNCTIONS FROM ATOMIC ORBITALS

Let us now review the theoretical procedure by which an *accurate* many-electron state wave function may (at least in principle) be generated. Two points need to be noted here. First, both MO and VB approaches may be used in this process and, in the limit, both procedures ultimately lead to the same state wave function. Second, both MO and VB theories start off with the same basis set of AOs, but differ in the theoretical pathway taken. As we will discover, the two approaches differ in the *order* that particular procedures are carried out. A ·schematic diagram that shows the relationship between MO and VB approaches is presented in Scheme 1.1.

1.7.1 The MO Procedure

Step 1. *Generating the Set of MOs.* The starting point for both MO and VB methods begins with the set of AOs $\phi_1, \phi_2, \phi_3, \cdots \phi_N$. In the MO procedure

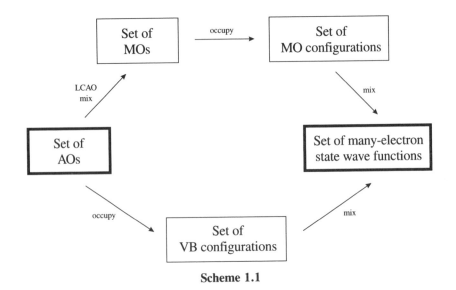

Scheme 1.1

(Scheme 1.1, upper pathway), AOs are mixed together by the LCAO procedure, to generate a set of MOs ψ_1, ψ_2, ψ_3, $\cdots$, ψ_N (we will discuss this procedure in detail in Chapter 2).

Step 2: *Generating the Set of MO Configurations.* Creation of the set of electronic configurations Ψ_0, Ψ_1, Ψ_2, $\cdots$, Ψ_N is brought about by allocating the electrons to the MOs. Each arrangement of electrons within the set of MOs generates a unique electronic configuration, as shown in Eqs. (1.9) and (1.10) for Ψ_0 and Ψ_1.

Step 3. *Generating the Set of Many-Electron State Wave Functions.* This final step has similarities to the first step, in which AOs were mixed to form MOs. In this step, however, *electronic configuration wave functions are mixed to form a set of electronic state wave functions.* This step is commonly termed *configuration interaction* (CI). For example, the ground-state wave function Ψ_g may be expressed as a linear combination of configurations.

$$\Psi_g = c_0\Psi_0 + c_1\Psi_1 + c_2\Psi_2 + \cdots c_N\Psi_N \qquad (1.20)$$

So even though a particular electronic configuration may be a useful representation of a particular state, it is not an *accurate* representation of that state. A more accurate wave function for a particular state can only be obtained through this procedure of configuration interaction. In similar fashion the range of excited-state wavefunctions Ψ_{e1}, Ψ_{e2}, Ψ_{e3}, and so on, may be obtained by taking *different* linear combinations of the same basis set of electronic configurations. This procedure in which configurations are allowed to interact will be discussed in more detail in Chapter 3.

In summary, we see that the MO procedure for generating a many-electron state wave function from AOs involves three steps: the LCAO mixing of AOs to generate MOs, occupying the MOs to generate a set of configurations, and finally, mixing the configurations to generate a set of states. The entire procedure is illustrated schematically in Scheme 1.1 (upper pathway).

1.7.2 The VB Procedure

In the VB procedure, the steps taken are essentially the same steps as in the MO procedure. However, as seen in Scheme 1.1 (lower pathway), the *order* in which the steps are carried out is different.

Step 1. *Generating the Set of VB Configurations.* As a first step the AOs are occupied to generate the range of VB configurations Ψ_0, Ψ_1, Ψ_2, $\cdots$, Ψ_L. These configurations, if represented in the form of structural formulas, are analogous to the resonance structures of classical organic chemistry. The mathematical form of a configuration wave function, for example, the ground configuration Ψ_0, is given by

$$\Psi_0 = (\phi_1)^2(\phi_2)^2(\phi_3)^2 \cdots (\phi_i)^2 \qquad (1.21)$$

where ϕ_1, ϕ_2, ϕ_3, $\cdots$, ϕ_i, is the set of AOs. Thus a VB configuration is simply a product function of the occupied AOs, and the set of VB configurations may be obtained by populating the AOs in different ways.

Step 2. *Generating the Set of Many-Electron State Wave Functions.* The second and final step in the VB procedure involves the mixing of the set of VB configurations to generate the many-electron state wave functions for the various states. This step is analogous to the CI step in the MO procedure (Step 3). Thus the ground state may be described by Ψ_g, as shown in Eq. (1.20) (though the configurations involved in the mixing procedure, Ψ_0, Ψ_1, Ψ_2, $\cdots$, Ψ_L, are now VB configurations). *If the VB configurations are represented in pictorial form (i.e., like resonance structures) rather than in mathematical form, then our description of the state resembles the common chemical procedure in which a species is described as a hybrid of several resonance structures.*

We stress again that the many-electron state wave function, obtained by either the VB or MO methods, is, in the limit, identical; the difference between the two procedures hinges on the *order* in which the electron occupancy and mixing procedures are carried out. So the question of which method is more correct is quite meaningless. Both methodologies, when carried out to their limits, lead to the same end point. However, we will discover in the following chapters that the *chemical* insight that can be obtained by following each of the two routes, is often different, so that each methodology has its special advantages.

1.7.3   State Wave Functions for the Hydrogen Molecule

Let us clarify the general scheme described above, in which state wave functions are built up from AOs (Scheme 1.1), by applying the procedure to a specific example: the hydrogen molecule. This is illustrated in Scheme 1.2.

In the MO method, LCAO mixing of the hydrogen *1s* orbitals generates unoccupied σ and σ^* orbitals. Allocating the two available electrons to the two orbitals in the different possible arrangements generates the various electronic configurations, [Eqs. (1.6–1.8)], which can then be mixed in different combinations to generate the state wave functions for the different electronic states of H_2. For example, the ground-state wave function of the hydrogen molecule Ψ_g is only approximately described by the ground configuration σ^2, and may be described more accurately by a wave function obtained by the configuration interaction of the ground configuration σ^2, with the excited configuration σ^{*2}, as shown in Eq. (1.22) (where λ is a mixing parameter between 0 and 1). Thus, Eq. (1.22) represents a better electronic description of the ground *state* of H_2 in terms of MO configurations (we will explain why in Section 1.7.4).

$$\Psi_g = \sigma^2 - \lambda(\sigma^{*2}) \tag{1.22}$$

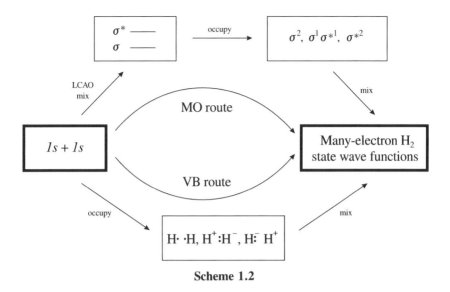

Scheme 1.2

In the VB method, the first step involves occupying the H $1s$ orbitals in the various possible ways so as to obtain the three possible VB configurations [Eqs. (1.14–1.16)]. Mixing of these three VB structures results in the generation of the set of state wave functions for H_2. This means that in VB terms the ground-state wave function Ψ_g of H_2 may be represented by

$$\Psi_g = C_1(H\cdot \ \cdot H) + C_2(H\mathord{:}^- \ H^+) + C_3(H^+ \ \mathord{:}H^-) \tag{1.23}$$

We pointed out that the resultant many-electron state wave functions obtained by these two procedure (MO and VB) are, in the limit, the same. Let us now demonstrate this equivalence, by converting the MO state wave function into its VB analog.

1.7.4 Equivalence of MO and VB State Wave Functions

If we substitute the AO description of the σ bond [Eq. (1.3)] into the MO ground configuration wave function of H_2 [(Eq. (1.6)], we obtain (leaving out the normalization constant for simplicity)

$$\sigma^2 = (\phi_A + \phi_B)^2 \tag{1.24}$$

or more explicitly,

$$\sigma^2 = (\phi_A + \phi_B)(1)(\phi_A + \phi_B)(2) \tag{1.25}$$

Multiplying out this expression and ignoring the spin functions, we obtain

$$\sigma^2 = \phi_A(1)\phi_B(2) + \phi_A(2)\phi_B(1) + \phi_A(1)\phi_A(2) + \phi_B(1)\phi_B(2) \quad (1.26)$$

If we now represent Eq. (1.26) in pictorial form, as discussed in Section 1.5.3, we obtain

$$\sigma^2 = (H\cdot \ \cdot H) + (H{:}^- \ H^+) + (H^+ \ {:}H^-) \quad (1.27)$$

where the first two terms of Eq. (1.26) are represented by the covalent configuration $(H\cdot \ \cdot H)$, and the remaining two terms are represented by the two ionic configurations $(H{:}^- \ H^+)$ and $(H^+ \ {:}H^-)$, respectively.

Thus we have converted the ground configuration MO wave function σ^2 into a VB type wave function simply by describing the MOs in terms of the contributing AOs. Furthermore, as a result of the conversion, we can now see more clearly that this MO ground configuration is an imprecise description of the ground *state*; it *overemphasizes* the ionic contribution [which, from Eq. (1.26), can be seen to be 50%] at the expense of the covalent character. Similarly, the VB ground configuration, described by Eq. (1.14), is also an imprecise description of the ground state because it totally *ignores* the ionic contributions to the wave function.

Let us now see how interaction of MO configuration σ^{*2} with σ^2 leads to a *better* description of the ground-state MO wave function Ψ_g. Just as we represented σ^2 in terms of AOs [Eq. (1.25)], we can do the same for σ^{*2}. Thus

$$\sigma^{*2} = (\phi_A - \phi_B)(1)(\phi_A - \phi_B)(2) \quad (1.28)$$

Multiplying out this expression we obtain

$$\sigma^{*2} = \phi_A(1)\phi_A(2) + \phi_B(1)\phi_B(2) - \phi_A(1)\phi_B(2) - \phi_A(2)\phi_B(1) \quad (1.29)$$

or pictorially

$$\sigma^{*2} = (H{:}^- \ H^+) + (H^+ \ {:}H^-) - (H\cdot \ \cdot H) \quad (1.30)$$

The primary difference between σ^2 and σ^{*2} is in the *sign* of the covalent contribution to the wave function—positive in the former, negative in the latter. So by mixing in of σ^{*2} into σ^2 with a *negative* sign, as indicated in Eq. (1.22), we *increase* the weight of the covalent contribution, while simultaneously *decreasing* the weight of the ionic contributions. This mathematical manipulation leads to the result we want—a more accurate description of the ground-state molecule.

How much mixing of σ^{*2} is required? The extent of this mixing, governed by the parameter λ is varied so as to optimize the energy of the system. So we see that despite the superficial distinction between the two ground-state descriptions of H_2, as expressed in Eqs. (1.22) and (1.23), the two are theoretically *equivalent* and readily interconverted.

REFERENCES

1. W. Heitler and F. London, *Z. Phys.* **44,** 455 (1927).
2. L. Pauling, *The Nature of the Chemical Bond*, Cornell University Press, Ithaca, New York, 1939.
3. F. Hund, *Z. Phys.* **51,** 759 (1928); **73,** 1 (1931).
4. R. S. Mulliken, *Phys. Rev.* **32,** 186 (1928); **41,** 49 (1932); *J. Chem. Phys.* **1,** 492 (1933).
5. E. Hückel, *Z. Phys.* **60,** 423 (1930); **72,** 310 (1931); *Trans. Faraday Soc.* **30,** 40 (1934).
6. L. Pauling and E. B. Wilson, Jr., *Introduction to Quantum Mechanics*, McGraw-Hill Book Co., New York, 1935.
7. J. N. Murrell, S. F. A. Kettle, and J. M. Tedder, *The Chemical Bond*, Wiley, New York, 1978.
8. P. A. M. Dirac, *Proc. R. Soc. (London)* **123,** 714 (1929).

2

MOLECULAR ORBITAL THEORY

In Section 1.4.1 we described the formation of the covalent bond in the hydrogen molecule in MO terms, and in Section 1.7.1 we outlined the general framework for the building up of a state wave function by the MO procedure. In this chapter we will discuss the MO approach in more detail, focusing on the first step in the MO procedure—the building up of MOs from AOs. We will examine this procedure both (a) *mathematically* and (b) *qualitatively*, since both methodologies play a key role in modern organic chemistry.

The mathematical formalism for building up MOs provides the basis for the computational procedures (Hückel, semiempirical, and ab initio), whereby organic structures and reaction surfaces may be effectively calculated. The fact that real structures and reaction surfaces can now be studied without recourse to the reagent bottle, the test tube, or even the spectrometer, represents nothing less than a revolution in the way that modern chemistry is performed; chemical knowledge is now accessible, not only by carrying out experiments in the laboratory, but also by carrying out calculations on a computer. So though it might sound paradoxical, "computational chemistry" is able to provide the chemist with an important new source of chemical "data." Indeed, there are well-documented cases where the computational data have been known to improve on, or even invalidate, experimental findings (obviously incorrectly obtained ones!) so that the impact and importance of the computational procedures are now universally recognized.[1]

However, despite their great utility, computationl technology has not satisfied the chemist's need to understand. The mathematical complexity of the computational procedure is such that the qualitative reasoning behind a particular result is often lost within the mathematical labyrinth. For this reason *qualitative* theory fulfills an important role: To provide *insight* into chemical phenomena so that the chemist can *understand* a particular experimental result, so he/she can explain *why* a particular calculation came out the way it did, and to enable him/her to make *predictions* regarding unknown phenomena (without recourse to the laboratory *or* the computer). In this chapter we will describe both quantitative and qualitative MO theory and note how each has made important contributions to our understanding of organic structure and reactivity.

2.1 QUANTITATIVE MOLECULAR ORBITAL THEORY

The main goal of MO calculations is to be able to quantify molecular properties, most commonly structure and energy, by solving the Schrödinger equation (1.2). Since an exact solution is unobtainable for a system with more than one electron, and because practical chemistry involves many-electron systems, MO theory relies heavily on the use of certain approximations and assumptions. The most important assumptions relate to the form of the state wave function Ψ and constitute an integral part of the LCAO method.

2.1.1 Linear Combination of Atomic Orbitals Method

At the simplest level of approximation it is assumed that the *state* wave function may be usefully approximated by a *configuration* wave function. This key assumption not only simplifies the MO procedure mathematically, but, as we will see in Section 2.2, also enables the MO procedure to be applied in a qualitative and pictorial fashion. This occurs because the configuration description of the state wave function provides a *physical* picture of the way electrons are distributed within a molecule by specifying which orbitals are occupied by electrons. According to this picture *the nuclei of a molecule can be considered as enveloped by a set of MOs, some occupied by electrons, some unoccupied, and that the structural and reactivity characteristics of molecules may be understood by considering the energy, geometry, and symmetry properties of these MOs.* (Unfortunately, once the simple *single configuration* wave function is abandoned for more mathematically precise *multiconfiguration* wave functions, the pictorial simplicity of the MO description is lost; MOs, as a physical description of where electrons are located within a molecule, only "exist" at the single configuration level of theory.)

A second important assumption that governs the way MOs are constructed is that a set of MOs may be built up from the set of AOs (ϕ_1, ϕ_2, . . . , ϕ_j, . . . , ϕ_N) associated with the atoms in the system. This procedure is termed the *linear combination of atomic orbitals* or LCAO method, since this is the way the MOs are expressed mathematically, as is shown in Eq. (2.1)

$$\psi_1 = c_{11}\phi_1 + c_{12}\phi_2 + c_{13}\phi_3 + \cdots c_{1N}\phi_N$$

$$\psi_2 = c_{21}\phi_1 + c_{22}\phi_2 + c_{23}\phi_3 + \cdots c_{2N}\phi_N$$

$$\vdots$$

$$\psi_N = c_{N1}\phi_1 + c_{N2}\phi_2 + c_{N3}\phi_3 + \cdots c_{NN}\phi_N \qquad (2.1)$$

or more concisely as

$$\psi_i = \sum_{j=1}^{N} c_{ij}\phi_j \qquad (2.2)$$

where c_{ij} are the coefficients that specify the contribution of a given AO, ϕ_j, to a particular MO, ψ_i. Thus we see that because the set of AOs may be combined in different ways, the LCAO procedure generates a *set of MOs*, and each MO is defined by a different set of AO coefficients. An important feature of this mixing process is that the number of MOs generated, N, always equals the number of AOs from which the MOs are built. A schematic representation of a set of MOs, ranked according to their energies, was shown in Fig. 1.2.

Some intuitive justification for the LCAO procedure can be proposed. First, when a molecule is separated into its constituent atoms, it is clear that the MOs must evolve smoothly into the constituent AOs. So this intrinsic AO character must be represented explicitly within the MO description. Second, in the molecule an electron that is close to a particular atomic nucleus at a given moment will respond primarily to the potential of that nucleus, which is, of course, defined by the AOs. So, describing an MO as a linear combination of AOs seems physically reasonable.

In order to specify the MOs for a particular system we need to know the AO wave functions and the magnitudes of the coefficients c_{ij}. The AO functions are predetermined, but the coefficients that effectively define the set of MOs need to be evaluated for each individual system. How can this be done? In order to understand the mathematical procedure by which the sets of coefficients are determined, we need to consider an important principle of quantum chemistry: *the variation principle*.

2.1.2   The Variation Principle

To solve the Schrödinger equation for a particular system, we need to find the state wave function Ψ_{true}, for that system. Since this wave function is unknown, we replace it by some approximate *trial* eigenfunction Ψ_{approx}. But solving the Schrödinger equation with the trial eigenfunction Ψ_{approx} will yield a value for the energy E_{approx} that differs from the true energy of the system E_{true}. This result, in itself, is not very useful. However, the variation principle provides us with an additional piece of information that is crucial. It informs us that the calculated energy of the system E_{approx} *will always be greater than, or equal to, the true energy* E_{true}, as expressed in Eq. (2.3). So the better our approximation for the state wave function, the *lower* the energy calculated from the Schrödinger equation, and the closer it will approach the true energy.

$$E_{approx} \geq E_{true} \tag{2.3}$$

Formal proof of the variation principle will not be presented here.[2] We will satisfy ourselves with a simple pictorial analogy that may make the idea more intuitive. Consider a student studying for an exam; the more he/she studies the material to be examined, the fewer the errors he/she will make. Of course, the very best a student can do is to achieve a perfect result with no errors at all! In this analogy the true energy, obtained with the exact wave function, is equivalent to this perfect exam. Solving the Schrödinger equation with an approximate wave function *cannot* yield a lower energy than the one that would be obtained with the exact (though unknown) wave function—nature has done the best job possible!

How does the variation principle assist us in obtaining a good approximation for the state wave function? We have already determined that the LCAO procedure will form the basis for generating the set of MOs from which a configuration wave function (our trial eigenfunction) can be constructed. So, to obtain the set of MOs [Eq. (2.1)] from the basis set of AOs, we need to calculate the values of the AO coefficients c_{ij}. The variation principle tells us *that we need to choose these coefficients in a way that will minimize the energy of the configuration.* Let us see how this is done.

We begin by expressing the Schrödinger equation with energy as the subject. To do this we first multiply both sides of the Schrödinger equation by Ψ to give

$$\Psi \mathcal{H} \Psi = \Psi E \Psi \tag{2.4}$$

which can be simplified to

$$\Psi \mathcal{H} \Psi = E \Psi^2 \tag{2.5}$$

(the Ψ functions on the left-hand side cannot be combined since $\mathcal{H}$ is a mathematical operator that operates on just one of the Ψ values). Integrating over all space, and isolating E leads to

$$E = \frac{\displaystyle\int \Psi \mathcal{H} \Psi \, d\tau}{\displaystyle\int \Psi^2 \, d\tau} \tag{2.6}$$

So Eq. (2.6) is just another form of the Schrödinger equation, with E as the subject.

Let us now express Eq. (2.6) in terms of the set of AO coefficients ϕ_j and their coefficients c_i. Substituting Eq. (2.2) into Eq. (2.6) leads to

$$
E = \frac{\int \left(\sum_i c_i \phi_i \right) \mathcal{H} \left(\sum_j c_j \phi_j \right) d\tau}{\int \left(\sum_i c_i \phi_i \right) \left(\sum_j c_j \phi_j \right) d\tau}
\tag{2.7a}
$$

which can be rearranged to

$$
E = \frac{\sum_i \sum_j c_i c_j \int \phi_i \mathcal{H} \phi_j \, d\tau}{\sum_i \sum_j c_i c_j \int \phi_i \phi_j \, d\tau}
\tag{2.7b}
$$

where both i and j are now independent AO counters [i is no longer the MO subscript index, as used in Eq. (2.2), and has been dropped for simplicity].

The form of Eq. (2.7) is cumbersome and can be simplified by the following representations:

$$
H_{ij} = \int \phi_i \mathcal{H} \phi_j \, d\tau
\tag{2.8}
$$

$$
S_{ij} = \int \phi_i \phi_j \, d\tau
\tag{2.9}
$$

Thus Eq. (2.7) now may be written more succinctly as

$$
E = \frac{\sum_i \sum_j c_i c_j H_{ij}}{\sum_i \sum_j c_i c_j S_{ij}}
\tag{2.10}
$$

According to the variation principle, we now need to minimize the value of E by the appropriate choice of AO coefficients c. Mathematically this can be done by differentiating the energy with respect to each of the c values as follows:

$$
\frac{\partial E}{\partial c_i} = 0
\tag{2.11}
$$

where c_i is the family of coefficients for each of the AOs. This leads to the set of N equations

$$c_1(H_{11} - ES_{11}) + c_2(H_{12} - ES_{12}) \cdots c_N(H_{1N} - ES_{1N}) = 0$$
$$c_1(H_{21} - ES_{21}) + c_2(H_{22} - ES_{22}) \cdots c_N(H_{2N} - ES_{2N}) = 0$$
$$\vdots \qquad\qquad \vdots \qquad\qquad \vdots \qquad\qquad (2.12)$$
$$c_1(H_{N1} - ES_{N1}) + c_2(H_{N2} - ES_{N2}) \cdots c_N(H_{NN} - ES_{NN}) = 0$$

whose solution may be obtained by solving the following determinant [Eq. (2.13)], termed the *secular determinant.*

$$\begin{vmatrix} H_{11} - ES_{11} & H_{12} - ES_{12} & \cdots & H_{1N} - ES_{1N} \\ H_{21} - ES_{21} & H_{22} - ES_{22} & \cdots & H_{2N} - ES_{2N} \\ \vdots & \vdots & \vdots & \vdots \\ H_{N1} - ES_{N1} & H_{N2} - ES_{N2} & \cdots & H_{NN} - ES_{NN} \end{vmatrix} = 0 \qquad (2.13)$$

To solve this determinant for the set of MOs and their corresponding energies, the magnitudes of the matrix elements H_{ij} and S_{ij} need to be calculated. Ab initio calculations calculate these elements with no assumptions or empirical input, while semiempirical methods simplify the procedure by assuming certain integrals to be zero, and use empirical data to avoid having to calculate others. Of course there are different ways of incorporating these kinds of simplifications. This is the reason there are different semiempirical procedures. One of the earliest of these was the *complete neglect of differential overlap* (CNDO) procedure, which assumes the AO basis functions do not overlap. Subsequently, methods such as *intermediate neglect of differential overlap* (INDO), *modified neglect of differential overlap* (MNDO), *modified intermediate neglect of differential overlap* (MINDO), *Austin Model 1* (AM1), *parameterized method 3* (PM3), and most recently, *Semi-Ab-initio Model 1* (SAM1), with less drastic assumptions, have been developed.[3] While the relative merits of the two approaches, ab initio and semiempirical, have raised passions among computational chemists, the more fundamental ab initio approach appears to have won the wider following.

2.1.3 Significance of ψ and the Normalization Condition

Whereas the wave function ψ is a mathematical function whose direct physical significance is not apparent, the square of the wave function (ψ^2) can be given a direct physical interpretation: *the probability distribution of finding the elec-*

tron at any given point in space. In other words, a diagram of ψ^2 for a particular wave function represents an *electron density picture* of that particular orbital when occupied by an electron. Since the total probability of finding an electron in an orbital must sum up to unity, we can write

$$\int \psi^2 \, d\tau = 1 \tag{2.14}$$

Equation (2.14) is termed the *normalization condition* and when it holds we say the wave function is *normalized*. For any MO, expressed as a linear combination of (normalized) AOs,

$$\psi = c_1\phi_1 + c_2\phi_2 + c_3\phi_3 + \cdots + c_N\phi_N \tag{2.15}$$

and making the simplifying assumption that $S_{ij} = 0$, the normalization condition becomes

$$c_1^2 + c_2^2 + c_3^2 + \cdots + c_N^2 = 1 \tag{2.16}$$

Thus, we can write the normalized MO wave function as

$$\psi = \frac{1}{N} \{c_1\phi_1 + c_2\phi_2 + c_3\phi_3 + \cdots + c_N\phi_N\} \tag{2.17}$$

where

$$N = \sqrt{c_1^2 + c_2^2 + c_3^2 + \cdots + c_N^2} \tag{2.18}$$

We will find Eq. (2.18) valuable in calculating Hückel MO coefficients.

2.1.4 Hückel Method

As noted earlier, simplifying assumptions play a major role in all computational procedures, and the Hückel method takes this process to the extreme. Despite this, however, it is of special importance for the insights it provides into the chemistry of π systems. For this reason we will consider the Hückel method in some detail.

We have seen that solving the secular determinant is the key to obtaining the set of MOs from a basis set of AOs. Since evaluation of the integrals H_{ij} and S_{ij} is complex and often demands considerable computing time, solution of the secular determinant can be simplified enormously if the values of these integrals can be incorporated as parameters, thereby avoiding the need to calculate them at all. This is the essence of the Hückel method. While simplifying the procedure in this way does limit its utility as a quantitative method, it does make the mathematical procedure more "transparent" and this is an aspect of great importance. Indeed, Hückel calculations on simple systems are suited to a "back-of-the-envelope" analysis. Furthermore, we will later see that the Hückel method is a useful accompaniment to the qualitative approaches discussed in Section 2.2.

The Hückel method is based on the following limitations and assumptions:

1. *The method is applied specifically to the analysis of conjugated π systems.* The Hückel method provides information regarding the MO energy levels of conjugated π systems, and is not used for analyzing the more complex σ framework. Accordingly, the basis set of AOs used to build up the MOs are the set of atomic $2p$ orbitals on each of the carbon atoms of the conjugated system. Other AOs associated with the underlying σ framework are not considered. So the π systems of molecules, such as benzene, butadiene, allyl cation and anion, are typical of the electronic systems that can be analyzed using the Hückel method.

2. *Values of the energy integrals H_{ij} are parameterized.* When $i = j$, the integral H_{ii} is just the energy of an electron in the $2p$ AO of a C atom. It is assumed to be the same for all the C atoms that contribute to the conjugated system and is given the value α (~ 10 eV—the ionization potential of the methyl radical). Thus,

$$H_{ii} = \alpha \qquad (2.19)$$

The sign of α is negative since the energy of an electron in *any* AO is *lower* than that of a free electron, whose energy is defined as zero. The parameter H_{ii} is commonly termed the *coulomb integral* since its magnitude is primarily governed by the coulomb forces (nuclear–electron and electron–electron).

3. The value of H_{ij} ($i \neq j$) depends on whether or not the two carbon atoms, i and j, are directly bonded to one another. *If the two C atoms are bonded to one another, the value of H_{ij} is given the value β and assumed to be the same for any two neighboring C atoms.* Thus,

$$H_{ij} = \beta \qquad (2.20)$$

The parameter β is more difficult to describe in physical terms, but may be thought of as the energy of the electron in the overlap region of the two AOs, since the integral of Eq. (2.20) involves both the orbitals ϕ_i and ϕ_j. The parameter β is commonly termed the *resonance integral* since the stabilization energy that arises from the interaction between ϕ_i and ϕ_j may be considered as due to the resonance of the electron between these two orbitals. The sign of β, like α, is negative. If the two C atoms are not bonded directly to one another, then H_{ij} is presumed equal to zero.

4. $S_{ii} = 1$ (by definition, since this is just the normalization condition), but the value of S_{ij} is assumed to be 0, even for neighboring C atoms.

With the above assumptions, the secular determinant [Eq. (2.13)] takes on the following form (shown arbitrarily for a five-carbon system):

$$\begin{vmatrix} \alpha - E & ? & ? & ? & ? \\ ? & \alpha - E & ? & ? & ? \\ ? & ? & \alpha - E & ? & ? \\ ? & ? & ? & \alpha - E & ? \\ ? & ? & ? & ? & \alpha - E \end{vmatrix} = 0 \qquad (2.21)$$

The diagonal elements all become $\alpha - E$ since H_{ii} equals α and S_{ii} equals 1. The symbol ? stands for either β or 0, depending on whether the two atoms i and j, defined by the matrix element, are adjacent to one another (in which case $H_{ij} = \beta$), or, are not adjacent (in which case $H_{ij} = 0$).

To simplify the algebra further, it is useful to divide the secular determinant throughout by β and to replace the diagonal terms that result, $(\alpha - E)/\beta$, by x according to Eq. (2.22).

$$x = (\alpha - E)/\beta \qquad (2.22)$$

Following this substitution the secular determinant becomes

$$\begin{vmatrix} x & ? & ? & ? & ? \\ ? & x & ? & ? & ? \\ ? & ? & x & ? & ? \\ ? & ? & ? & x & ? \\ ? & ? & ? & ? & x \end{vmatrix} = 0 \qquad (2.23)$$

where the symbol ? now needs to be replaced by either 1 or 0, depending on whether the two carbon atoms defined by the particular off-diagonal matrix element are adjacent or not [in analogy with Eq. (2.21)].

It is easiest to understand the Hückel method with the aid of some examples, so let us begin with the simplest of all π systems: the ethylene molecule.

2.1.4A   Ethylene. For the π system of ethylene the secular determinant is just

$$\begin{vmatrix} x & 1 \\ 1 & x \end{vmatrix} = 0$$

This leads to

$$x^2 - 1 = 0$$

whose roots are

$$x = \pm 1$$

The two roots indicate that two MOs are formed. Since $E = \alpha - x\beta$ [a rearranged form of Eq. (2.22)], the orbital energies of the two MOs become

$$E_1 = \alpha + \beta \qquad \text{(the low-energy orbital)} \qquad (2.24)$$

$$E_2 = \alpha - \beta \qquad \text{(the high-energy orbital)} \qquad (2.25)$$

Let us now determine the MO coefficients for these orbitals so we can write down their wave functions.

For ethylene, the secular equations (2.12) are

$$c_1(\alpha - E) + c_2\beta = 0 \qquad (2.26)$$

$$c_1\beta + c_2(\alpha - E) = 0 \qquad (2.27)$$

Rearranging Eq. (2.26) we may write

$$c_1/c_2 = -\beta/(\alpha - E)$$

Substituting $(\alpha + \beta)$ for E (the value of E_1) we obtain

$$c_1/c_2 = 1$$

that is,

$$c_1 = c_2$$

Given the normalization condition [$N = (1^2 + 1^2)^{1/2}$ from Eq. (2.18)], we can now specify the values of c_1 and c_2 as

$$c_1 = c_2 = 1/\sqrt{2}$$

Thus the wave function of the low-energy MO is

$$\psi_1 = \frac{1}{\sqrt{2}}(\phi_1 + \phi_2) \qquad (2.28)$$

Similarly, by substituting the value of E_2 into Eqs. (2.26) or (2.27) we obtain the wave function of the high-energy MO as

$$\psi_2 = \frac{1}{\sqrt{2}}(\phi_1 - \phi_2) \qquad (2.29)$$

2.1.4B Allyl. The allyl system is built up from three $2p$ AOs, so the secular determinant is given by

$$\begin{vmatrix} x & 1 & 0 \\ 1 & x & 1 \\ 0 & 1 & x \end{vmatrix} = 0 \qquad (2.30)$$

The two zeros in the determinant (at the 1,3 and 3,1 positions in the matrix) reflect the fact that there is no bonding between C1 and C3, while the 1's (at the 1,2; 2,1; 2,3; and 3,2 positions), reflect the fact that C1 and C2, and C2 and C3, *are* bonded together.

Solving the determinant leads to

$$x^3 - 2x = 0$$

whose solutions are $x = -\sqrt{2}$, 0, and $+\sqrt{2}$
So mixing three AOs leads to the formation of three MOs and from Eq. (2.22) we obtain their energies as

$$E_3 = \alpha - \sqrt{2}\beta$$

$$E_2 = \alpha$$

$$E_1 = \alpha + \sqrt{2}\beta$$

To obtain the coefficients of the MOs we need to solve the variation equations (from which the secular determinant was derived). These are just

$$c_1 \cdot x + c_2 \cdot 1 + c_3 \cdot 0 = 0 \qquad (2.31a)$$

$$c_1 \cdot 1 + c_2 \cdot x + c_3 \cdot 1 = 0 \qquad (2.31b)$$

$$c_1 \cdot 0 + c_2 \cdot 1 + c_3 \cdot x = 0 \qquad (2.31c)$$

For $x = -\sqrt{2}$, the lowest energy orbital, solution of these equations yields

$$c_2 = \sqrt{2}c_1 \qquad (2.32)$$

$$c_1 = c_3 \qquad (2.33)$$

Since the normalization condition means that

$$c_1^2 + c_2^2 + c_3^2 = 1 \qquad (2.34)$$

we find from Eqs. (2.32–2.34) that

$$c_1 = \tfrac{1}{2}$$

$$c_2 = 1/\sqrt{2}$$

$$c_3 = \tfrac{1}{2} \tag{2.35}$$

Thus the lowest energy MO may be written as

$$\psi_1 = \frac{1}{2}\phi_1 + \frac{1}{\sqrt{2}}\phi_2 + \frac{1}{2}\phi_3 \tag{2.36}$$

Substituting $x = 0$ and $+\sqrt{2}$ into the Eqs. 2.31 leads to the coefficients for the other two MOs. These two MOs may be shown to be

$$\psi_2 = \frac{1}{\sqrt{2}}\phi_1 - \frac{1}{\sqrt{2}}\phi_3 \tag{2.37}$$

$$\psi_3 = \frac{1}{2}\phi_1 - \frac{1}{\sqrt{2}}\phi_2 + \frac{1}{2}\phi_3 \tag{2.38}$$

A pictorial representation of the 3 MOs for the allyl system are illustrated in Fig. 2.1. Note that ψ_2 is termed a nonbonding orbital since its energy α is identical to that of the $2p$ AO.

The orbital picture for the allyl system does not inform us about the electron occupancy of the three orbitals. In fact three species (the allyl cation, allyl radical, and allyl anion) may be built up from this set of three MOs. The orbital occupancy for these three species is illustrated in Fig. 2.2. The principle that

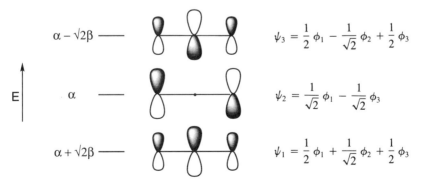

Fig. 2.1. Set of MOs for the allyl system showing the Hückel MO energies and their orbital coefficients.

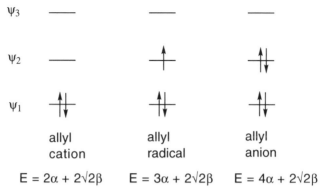

Fig. 2.2. The MO representations for the allyl cation, radical, and anion showing orbital occupancy and π-electron energies.

guides the way MOs are occupied is straightforward: For ground-state species (in contrast to excited-state species), MOs are filled in order of increasing energy, and each MO is filled with two electrons before the next highest MO is occupied. Thus, in the allyl cation, with just two π electrons, the two electrons are allocated to the lowest energy MO, ψ_1. In the three-electron allyl radical system, the third electron is placed into the next highest orbital, ψ_2. The four-electron system, the allyl anion, has two electrons in ψ_1 and two electrons in ψ_2.

Once the electron occupancy of a system of π-MOs is established, the *π-electron energy* and the *π-bond energy* may be evaluated. The π-electron energy is defined by

$$E_\pi = \sum_i n_i E_i \qquad (2.39)$$

where n_i is the number of electrons in the ith MO, and E_i is the energy of the ith MO. Thus E_π represents the *total* energy of the electrons in the π system. The *π-bond energy* (which should not be confused with the π-electron energy) is the bonding energy that results from the interaction of the π electrons. Hence, it is just the difference between the π-electron energy and the energy of the those electrons in nonbonding C $2p$ AOs (equal to $n\alpha$).

The π-electron bond order p_{jk}, which quantifies the extent of double-bond character between two adjacent carbon atoms j and k, is given by

$$p_{jk} = \sum_i n_i c_{ij} c_{ik} \qquad (2.40)$$

where c_{ij} and c_{ik} are the coefficients of the AOs, ϕ_j and ϕ_k, in the MO, ψ_i.

The π-electron charge density at atom j, labeled q_j, is given by

$$q_j = \sum_i n_i c_{ij}^2 \qquad (2.41)$$

from which the actual charge on that atom may be calculated. Let us demonstrate these ideas for the allyl cation.

The π-electron energy E_π for the two-electron allyl cation system may be found by substituting the MO energy ($\alpha + \sqrt{2}\beta$) and the number of electrons (two) into Eq. (2.39). This leads to a value of $2\alpha + 2\sqrt{2}\beta$.

The π-bond energy of the allyl cation may be found by subtracting the nonbonding energy of the two electrons (2α) from this total π-energy value to give a value of $2\sqrt{2}\beta$. This figure represents the bonding energy associated with the π system. It may be compared to the energy of the localized π bond in ethylene, which can be shown to be 2β (using the same procedure for ethylene). The difference between these two values of 0.82β, may be thought of as a measure of the *resonance* or *delocalization* energy in the allyl cation; that is, the *enhanced* stability of a two-electron π bond spread over *three* C atoms compared to a two-electron π bond localized on just two C atoms.

The π-bond order between C1 and C2 in allyl may be obtained from Eq. (2.40). Thus

$$p_{12} = 2 \cdot 1/2 \cdot 1/\sqrt{2} = 0.707$$

which compares with the bond order of 1 calculated for ethylene. In other words the double-bond character between two adjacent carbon atoms in the allyl cation is less than that in ethylene.

Finally, the π-electron charge density on C1 (as well as C3, by symmetry) is calculated to be $2\left(\frac{1}{2}\right)^2 = \frac{1}{2}$. Since the charge density of a nonbonding C atom is 1 (due to the single electron that each carbon contributes to the π system), C1 is found to carry a 0.5 positive charge (the difference between the π-electron charge density of 0.5 and the charge density of 1 on a nonbonding C). A similar calculation for the π-electron charge density on C2 gives $2(1/\sqrt{2})^2 = 1$, indicating zero charge. Thus these calculations suggest that the positive charge in the allyl cation is split equally between the C1 and C3 carbon atoms.

It is instructive to compare the MO description of the allyl cation with the corresponding VB description. In the VB description the additional stability of the π system, associated with the delocalized nature of the two π electrons, is represented by the interaction of the two resonance structures

$$CH_2{=}CH{-}CH_2^+ \leftrightarrow {}^+CH_2{-}CH{=}CH_2$$

Thus, in agreement with the MO analysis, the simple VB picture suggests (a) that there will be partial (and not full) double-bond character between adjacent carbon atoms, (b) that the positive charge will be divided equally between the two end carbon atoms, and (c) that the delocalization of the two electron π bond over three centers will lead to added stability—the so-called resonance stabilization. For the allyl cation at least, the qualitative descriptions provided by MO and VB approaches turn out to be quite similar.

2.1.4C Butadiene. The secular determinant for butadiene, a four-carbon system, is described by a 4×4 determinant

$$\begin{vmatrix} x & 1 & 0 & 0 \\ 1 & x & 1 & 0 \\ 0 & 1 & x & 1 \\ 0 & 0 & 1 & x \end{vmatrix} = 0 \qquad (2.42)$$

According to the method of cofactors, this 4×4 determinant can be broken down into smaller determinants.[4] Thus, Eq.(2.42) is reduced to just

$$x^4 - 3x^2 + 1 = 0$$

whose solutions are

$$x = \pm 1.62, \ \pm 0.62$$

Accordingly, this system of four AOs yields four MOs, and from Eq. (2.22), the four values of x yield the energies

$$E_4 = \alpha - 1.618\beta$$

$$E_3 = \alpha - 0.618\beta$$

$$E_2 = \alpha + 0.618\beta$$

$$E_1 = \alpha + 1.618\beta \qquad (2.43)$$

Solving for the coefficients leads to[4]

$$\psi_4 = 0.372\phi_1 - 0.602\phi_2 + 0.602\phi_3 - 0.372\phi_4$$

$$\psi_3 = 0.602\phi_1 - 0.372\phi_2 - 0.372\phi_3 + 0.602\phi_4$$

$$\psi_2 = 0.602\phi_1 + 0.372\phi_2 - 0.372\phi_3 - 0.602\phi_4$$

$$\psi_1 = 0.372\phi_1 + 0.602\phi_2 + 0.602\phi_3 + 0.372\phi_4 \qquad (2.44)$$

From the energies of the MOs, the (ground state) π-electron energy of butadiene is readily shown to be $4\alpha + 4.472\beta$, and hence the π-bond energy is 4.472β. The resonance stabilization of the conjugated butadiene system compared to two *unconjugated* π bonds as reference (as in two ethylene molecules), is obtained by subtracting the π-electron energy of two ethylene molecules, equal to $4(\alpha + \beta)$, from that of butadiene. Thus the resonance stabilization of butadiene becomes

$$4\alpha + 4.472\beta - 4(\alpha + \beta) = 0.472\beta$$

Pictorially, the set of four MOs may be illustrated as shown in Fig. 2.3.

Two points need to be noted. First, the *sizes* of the AOs in Fig. 2.3 are intended to represent the relative *magnitudes* of the coefficients of the respective AOs. Large orbitals in the pictorial representation correspond to large coefficients of the AOs in the MOs. Second, the orbital *phase* is a feature of great significance since it determines the number of nodal planes in a particular MO, which in turn governs the energy of that MO. Thus, if in a given MO the AO coefficients for two adjacent carbon atoms are of opposite sign, this is reflected pictorially in an inversion of one orbital with respect to the other, and the

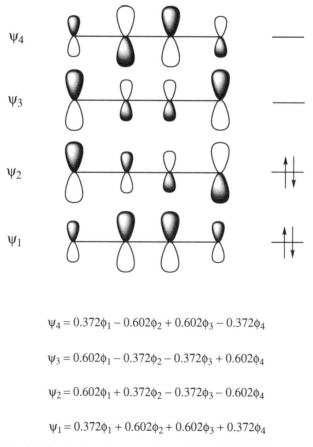

$$\psi_4 = 0.372\phi_1 - 0.602\phi_2 + 0.602\phi_3 - 0.372\phi_4$$

$$\psi_3 = 0.602\phi_1 - 0.372\phi_2 - 0.372\phi_3 + 0.602\phi_4$$

$$\psi_2 = 0.602\phi_1 + 0.372\phi_2 - 0.372\phi_3 - 0.602\phi_4$$

$$\psi_1 = 0.372\phi_1 + 0.602\phi_2 + 0.602\phi_3 + 0.372\phi_4$$

Fig. 2.3. Set of MOs for butadiene together with the Hückel MO wave functions.

presence of a nodal plane between the two AOs (see Fig. 2.3). Thus ψ_1, with no sign changes between adjacent AO coefficients, has no nodal planes, while ψ_4, with three sign changes between AO coefficients, has three nodal planes (between each pair of adjacent carbon atoms).

2.2 QUALITATIVE MOLECULAR ORBITAL THEORY

In Section 2.1 we described the way in which the LCAO concept may be applied quantitatively (through solution of the secular equations) to yield the wave functions for a set of MOs. At first sight the procedure appears to be essentially a mathematical one. However, the remarkable fact is that the process of building up MOs from AOs, for small molecules at least, is amenable to a simple qualitative procedure that requires no mathematical input at all. So MOs for molecules, such as benzene, butadiene, methane, and water, are quite accessible by a nonmathematical procedure. The pictorial tool by which MOs may be built up from AOs (or simpler MOs) is the *orbital interaction diagram*. However, before we proceed to discuss how orbital interaction diagrams are generated we need to discuss the physical significance of the MO, as suggested by Koopmans' theorem, and the role of *symmetry* in governing orbital interaction.

2.2.1 Ionization Potentials, Electron Affinities, and MO Energy Levels

Koopmans gave physical significance to the MO concept by arguing that the energy required to remove an electron from a molecule, its ionization potential, or ionization energy I, is simply the negative of the *highest occupied molecular orbital* (HOMO) energy (which is itself a negative number, so the sign of I is positive) from which the electron is ejected. Thus,

$$I = -\epsilon_{\text{HOMO}} \tag{2.45}$$

Since the energy required to remove an electron from a molecule can be obtained experimentally by techniques such as photoelectron spectroscopy, a direct link between the theoretically defined MO concept and a physical observable (the ionization potential) is established. In this way ionization potentials provide an experimental measure of orbital energies and may be usefully compared with the theoretically (i.e., computationally) obtained values. Ionization potentials for some common molecules are listed in Tables 7.1–7.3.

Actually, Koopmans' theorem tends to overestimate ionization potentials to some extent (often by $\sim 10\%$). No account is taken of the fact that orbital energies *after* ionization are not the same as those *before* ionization. Ionizing a molecule results in the MOs of the molecule, and especially the one from which the electron is removed, "relaxing" to lower energy, thereby effectively reducing the ionization energy. So Koopmans' theorem should not be viewed as a means of obtaining *accurate* orbital energies, and is probably best utilized as a *relative*, rather than an absolute, measure of orbital energy.

Just as Koopmans' theorem may be used to establish a relationship between the HOMO energy level and the ionization energy, the same idea can be used to relate the *lowest unoccupied molecular orbital* (LUMO) energy level to the *electron affinity A* of a molecule. Thus,

$$A = -\epsilon_{\text{LUMO}} \tag{2.46}$$

A more detailed discussion of ionization potentials and electron affinities appears in Chapter 7.

2.2.2 Symmetry

Symmetry arguments are extremely powerful in chemistry and in many cases reduce, what would otherwise be highly complex analyses, to dramatically simple ones. Thus a brief discussion of some elementary symmetry concepts is essential for a proper understanding of the way orbitals interact with one another.[5]

A *symmetry operation* is defined as an operation, which when performed on an object, results in a new orientation for the object that is *indistinguishable* from the original one. For example, rotating a circle 180° about a diameter is a symmetry operation because the circle, before rotation and after rotation, appear identical. Rotating a circle 180° about a tangent, however, would not constitute a symmetry operation: the arrangement, before and after rotation, *are* distinguishable; the circle has changed position.

Whenever a symmetry operation is carried out on an object, that operation is associated with some *symmetry element*. The important symmetry elements are

1. The *symmetry plane*, labeled σ, that bisects an object into two symmetrical halves.
2. The rotational axis, labeled C_n, where $360/n$ is the angle that the object is rotated through before it becomes indistinguishable from the original one (e.g., the axis of rotation of a circle about any diameter is a C_2 axis).
3. A center of symmetry i. A molecule possesses a center of symmetry, if, on proceeding in a straight line from any atom in the molecule through the center of symmetry, one meets the same atom on the other side of the center, at precisely the same distance from the center.

These ideas are easier to understand with the aid of some examples. Consider the molecules illustrated in Fig. 2.4.

The ammonia molecule possesses four symmetry elements. These are a C_3 *axis* along the y coordinate, which runs through the nitrogen atom (rotating the molecule by 120° leaves the molecule indistinguishable after the rotation), and three symmetry planes (labeled σ or σ_v—the v signifies the planes are vertical),

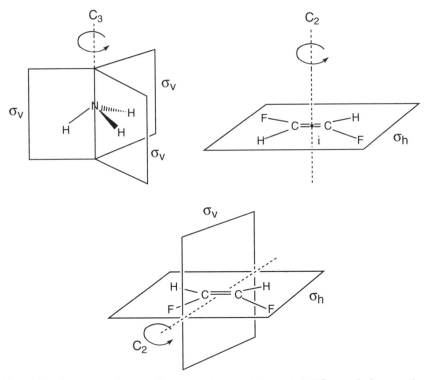

Fig. 2.4. Symmetry elements for ammonia (NH₃), *trans*-1,2-difluoroethylene, and *cis*-1,2-difluoroethylene.

each of which passes through one of the N—H bonds and the C_3 axis, bisecting the molecule into two symmetrical halves. *trans*-1,2-Difluoroethylene possesses three symmetry elements: a C_2 axis (perpendicular to the plane of the molecule and through the center of the double bond); a plane of symmetry σ_h (*h* stands for *horizontal*)—the plane of the molecule itself (in fact, *all planar molecules must exhibit this symmetry element*); and a center of symmetry *i*, at the center of the double bond. *cis*-1,2-Difluoroethylene also possesses three symmetry elements: A C_2 axis and two σ planes (the plane of the molecule and a perpendicular plane that bisects the double bond), but no center of symmetry.[a]

---

[a]Apart from the σ, C_n, and *i* elements, there exists a further symmetry element—the S_n axis, termed an improper axis, or rotation–reflection axis. This axis consists of rotation by 360°/*n*, followed by reflection in a plane perpendicular to the rotation axis. Allene, for example, possesses an S_4 axis. The S_n elements are less common than the preceding elements, though in saying this we do not consider S_n elements, where *n* = 1 and 2. Actually, a little thought will reveal that an S_1 axis is equivalent to a σ plane, while an S_2 axis is equivalent to a center of symmetry *i*. An interesting point is that the condition for chirality in a molecule may be defined in terms of symmetry elements. The condition for chirality is *that the molecule does not possess any S_n axis*. So a molecule might have some symmetry (e.g., a C_n axis) and still be chiral (e.g., *trans*-1,2-dichlorocyclobutane is chiral and has a C_2 axis), but it *cannot* have an S_n axis and still be chiral. For a more detailed description of symmetry see, for example, Ref. 5.

TABLE 2.1. Point Groups and Symmetry Elements for Some Common Molecules

Molecule	Point Group	Symmetry Elements
CHFClBr	C_1	None
1-Chloro-3-fluorobenzene	C_s	σ
trans-1,2-Difluorocyclobutane	C_2	C_2
trans-1,2-Difluoroethylene	C_{2h}	C_2, σ_h, i
H_2O	C_{2v}	C_2, $2\sigma_v$
NH_3	C_{3v}	C_3, $3\sigma_v$
Ethylene	D_{2h}	$3C_2$, 3σ, i
Methane	T_d	$4C_3$, $3S_4$, $3C_2$, 6σ
H—F	$C_{\infty v}$	C_{∞}, ∞C_v

Molecules that share the same set of symmetry elements are related in a symmetry sense and therefore need to be classified together. Such a set of molecules are said to belong to a particular *point group*. For example, molecules that possess *no* symmetry elements are said to belong to the C_1 group, while those, like *trans*-1,2-difluoroethylene, that possess the three elements C_2, σ_h, and i, are said to belong to the C_{2h} group. Table 2.1 illustrates a limited selection of molecules, the symmetry elements they possess, and the particular point group to which they belong. A more extensive table of point groups can be found in Ref. 5.

2.2.3   Orbital Symmetry

When symmetry arguments are applied to orbitals rather than to molecules, an additional factor must be considered: the phase characteristics of the orbitals. *An orbital may be classified as either symmetric or antisymmetric with respect to a given symmetry element.* This classification depends on the nodal properties of the orbital. Consider the *2p* orbital shown in **2.1**. This orbital possesses (among other symmetry elements) two symmetry planes, one horizontal (σ_h), along the x axis, and one vertical (σ_v), along the y axis. However, the orbital possesses a different symmetry character with respect to these two planes. Reflection of the orbital through the *horizontal* plane *switches* the positive and negative regions of the function, so the orbital is termed *antisymmetric* with respect to that plane. On the other hand, reflection of the orbital through the *vertical* plane leaves the phase distribution unchanged. So with respect to the vertical plane the orbital is classified as *symmetric*. Of course, a *1s* orbital, which has no nodes, is symmetric with respect to all of its symmetry elements.

The symmetry properties of orbitals are most important in determining whether two orbitals can overlap with one another *in principle*. Consider the approach of a *1s* orbital and a *2p* orbital, as indicated in Fig. 2.5. When the orbitals are in close proximity, regions of overlap will occur (indicated in Fig. 2.5a by the blacked out areas). However, because overlap is defined mathe-

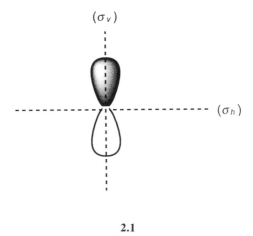

2.1

matically, overlap may either be *positive* or *negative* in sign. In the particular example shown in Fig. 2.5a, overlap in the region *above* the σ_h symmetry plane that bisects these two orbitals will be *positive* (overlap of a shaded area with a shaded area denotes positive overlap) while overlap *below* the plane will be *negative* (overlap of a shaded area with an unshaded area signifies negative overlap). From symmetry considerations, therefore, *the total overlap is zero—* the region of positive overlap precisely cancels out the region of negative overlap. Of course, if the two orbitals approach one another in a different orientation, as shown in Fig. 2.5b, then symmetry does *not* preclude overlap. As can be seen in the diagram, the two orbitals now do overlap.

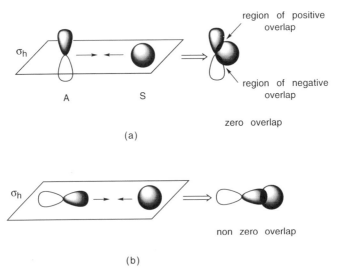

Fig. 2.5. Symmetry analysis of the overlap of *1s* and *2p* orbitals in two different orientations. (a) Orientation for zero overlap and (b) orientation for nonzero overlap.

The phenomenon of zero overlap and its symmetry cause has just been described in an intuitive way. However, the role of symmetry in governing orbital overlap can be defined more formally and may be expressed in the following rule: *Orbitals will overlap only if they are of the same symmetry (symmetric or antisymmetric) with respect to each of the symmetry elements shared by the system. A symmetry mismatch with respect to even one element will result in zero overlap.* Orbitals that do not overlap due to a symmetry mismatch are termed *orthogonal*. Let us apply this rule to the previous example of the interaction of a *2p* orbital with a *1s* orbital.

The system of a *2p* orbital and a *1s* orbital (shown in Fig. 2.5a) shares two common symmetry elements: a horizontal mirror plane (shown) and a vertical mirror plane (not shown). With respect to the horizontal plane, there is a symmetry mismatch; the *2p* orbital is *antisymmetric*, while the *1s* orbital is *symmetric* with respect to that plane. Consequently, these two orbitals are orthogonal and their overlap in this orientation is symmetry forbidden. In contrast, in Fig. 2.5b, where the orbital orientation is different, the two orbitals are now symmetric with respect to both horizontal and vertical symmetry planes, so the two orbitals *can* interact; that is, overlap in this orientation is symmetry allowed. (Note that the nodal plane of the *2p* orbital in Fig. 2.5b is *not* a symmetry element for the system of two orbitals in this orientation. While this plane *is* a symmetry element for the *2p* orbital itself, it is *not* a symmetry element for the *1s* orbital and so is not a symmetry element of the system as a whole. Only those symmetry elements common to *each* of the orbitals that constitute the system can be considered.)

In a similar manner, approach of two π orbitals face-on, as illustrated in Fig. 2.6a or b results in nonzero orbital overlap, (in this orientation both orbitals are symmetric with respect to the symmetry plane), while for the same approach of a π orbital and a π^* orbital, illustrated in Fig. 2.6c (where the π orbital is

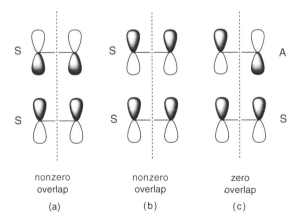

Fig. 2.6. Symmetry analysis of face-on π-π overlap and π-π^* overlap. (a) Nonzero overlap for head-to-head π-π overlap, (b) nonzero overlap for head-to-tail π-π overlap [identical to that in (a)], and (c) zero overlap for π-π^* overlap.

symmetric, and the π^* orbital is antisymmetric with respect to the mirror plane), there is a symmetry mismatch and overlap is zero.

Note that whether the two π orbitals are drawn in a head-to-head manner (Fig. 2.6a) or a head-to-tail manner (Fig. 2.6b) is arbitrary and of no importance. In the head-to-head orientation, the in-phase bonding MO that results from the mixing process will be obtained from the *positive* combination of the two MO functions, while the out-of-phase antibonding combination will result from the *negative* combination. In the case of a head-to-tail approach the reverse is true; the in-phase bonding MO is now obtained from the *negative* combination of the two MO functions. In other words it is unimportant whether interacting orbitals are drawn in a head-to-head manner (Fig. 2.6a) or a head-to-tail manner (Fig. 2.6b); the MOs that result from mixing are identical in both cases. The difference is taken care of mathematically by the appropriate positive or negative mixing sign. We will have more to say on the mixing procedure in Section 2.2.4.

2.2.4 Building Up Orbital Interaction Diagrams

Having discussed the role of symmetry in governing orbital overlap, we can now proceed to describe how to build up an orbital interaction diagram. The simplest interaction one can envisage is between two AOs. When two AOs interact with one another to form a pair of MOs, two situations need to be considered.

1. The two interacting orbitals are degenerate (i.e., of equal energy).
2. The two orbitals are nondegenerate.

Let us consider these in turn.

2.2.4A Degenerate Interactions. A degenerate interaction, exemplified by the mixing of two H $1s$ orbitals to form the two MOs of H_2 (discussed briefly in Section 1.4.1), is illustrated in Fig. 2.7. Interaction of two AOs, ϕ_A and ϕ_B, leads to the formation of two MOs: ψ_{MO}, the bonding combination, where the two orbitals mix in an in-phase manner and ψ^*_{MO}, the antibonding combination, where the two orbitals mix in an out-of-phase manner. The wave functions for these two orbitals are

$$\psi_{MO} = \frac{1}{\sqrt{2}} \, (\phi_A + \phi_B) \qquad (2.47)$$

$$\psi^*_{MO} = \frac{1}{\sqrt{2}} \, (\phi_A - \phi_B) \qquad (2.48)$$

Note the out-of-phase mixing leads to a nodal plane in the higher energy antibonding orbital.

Because the two AOs are degenerate, the contribution of each AO to each

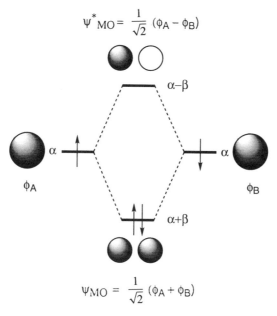

$$\psi^*_{MO} = \frac{1}{\sqrt{2}} (\phi_A - \phi_B)$$

$$\alpha - \beta$$

$$\alpha \qquad \phi_A \qquad \qquad \phi_B \qquad \alpha$$

$$\alpha + \beta$$

$$\psi_{MO} = \frac{1}{\sqrt{2}} (\phi_A + \phi_B)$$

Fig. 2.7. Orbital interaction diagram showing the degenerate interaction of two H $1s$ orbitals to form the two MOs of H_2. Here ψ_{MO} in the occupied bonding combination and ψ^*_{MO} is the unoccupied antibonding combination.

of the MOs, as reflected by the magnitudes of the AO coefficients, is *identical.* The fact that the coefficients are $1/\sqrt{2}$ (rather than 1), comes about from the need to *normalize* the wave function, as we discussed in Section 2.1.3. If we use the Hückel terminology and label the energies of the interacting AOs as equal to α, then the energy of the bonding MO equals $\alpha + \beta$, while that for the antibonding MO equals $\alpha - \beta$ [α and β are defined in Eqs. (2.19) and (2.20)]. Recall that α and β are *negative* in value since the zero reference point for the energy of an electron is defined as that of an unattached electron. It follows, of course, that the bonding MO is *stabilized* with respect to the AO energies (by an amount β), while the antibonding MO is *destabilized* with respect to the AO energies (also by the amount β).

The fact that *two* AOs interact to generate *two* MOs is significant. As noted in the quantitative discussion on the LCAO procedure, the number of MOs obtained through the AO mixing procedure is precisely the number of AOs from which the MOs are generated.

Effect of Electron Occupancy. We have seen how one can construct a pair of bonding and antibonding MOs from a pair of degenerate AOs. However, the chemical properties of the system will be greatly influenced by the number of electrons that occupy that set of MOs: the *electron occupancy* of the system. The system of two AOs may involve from one to four electrons and all four possibilities are illustrated in Fig. 2.8.

H_2^+

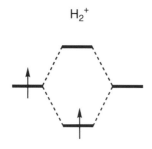

H_2

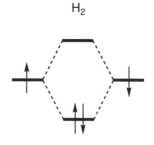

One-electron stabilization
R = 1.06 Å
Bond Energy = 61 kcal/mol

Two-electron stabilization
R = 0.74 Å
Bond Energy = 103 kcal/mol

He_2^+

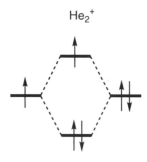

He_2

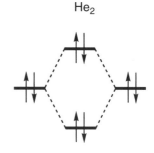

Three-electron stabilization
R = 1.08 Å
Bond Energy ≈ 50 kcal/mol

Nonexistent

Repulsive four electron interaction

Fig. 2.8. Effect of orbital occupancy (from 1–4 electrons) on the bond energy and bond length of an MO system comprising one bonding and one antibonding orbital (describing the species H_2^+, H_2, He_2^+, and He_2).

The one-electron situation is exemplified by the species H_2^+. The electron resides in ψ_{MO} and since just a single electron is involved in bonding, the bond is relatively weak (61 kcal mol^{-1}) and this is also reflected in the relatively long bond—1.06 Å. The two-electron situation (the hydrogen molecule) leads to a much stronger bond (103 kcal mol^{-1}), since now two electrons are stabilized by their being placed in ψ_{MO}. The two-electron bond strength is not twice the one-electron bond strength because the two-electron stabilization is partially canceled out by *electron–electron repulsion*; when two electrons occupy the same orbital, the magnitude of the electron–electron repulsion is relatively large since both electrons share the same physical space within the molecule.

Addition of a third electron to the system, as exemplified by the He_2^+ system, now *decreases* the overall bonding strength (down to ~ 50 kcal mol^{-1}). The drop in bond strength occurs because the third electron must be placed in an antibonding orbital ψ_{MO}^*, whose energy is *higher* than that of a noninteracting AO. In fact, the *antibonding* influence of the third electron approximately cancels the *bonding* influence of the second electron so that the overall bond strengths of the one-electron and three-electron systems turn out to be similar in magnitude.

An attempt to add four electrons to this system, as would be exemplified by the He_2 molecule (Fig. 2.8), leads to a nonbonding situation. Now the antibonding contribution of the two antibonding electrons in ψ_{MO}^* outweighs the bonding contribution of the two electrons in ψ_{MO}, so the He_2 molecule does not exist.

2.2.4B   Nondegenerate Interactions. The situation in which two orbitals of *different* energies interact is of fundamental importance because of its generality. It is formally treated by *perturbation theory*, since to first approximation each orbital represents a perturbation on the other. An energy diagram that demonstrates the interaction of two nondegenerate orbitals ϕ_A and ϕ_B is shown in Fig. 2.9. As in the case of the degenerate interaction discussed above, the two AOs interact to generate two MOs: one bonding (with no nodal plane) and one antibonding (with a single nodal plane). However, because the two orbitals

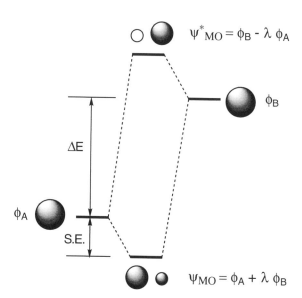

Fig. 2.9. Orbital interaction diagram showing the interaction of two nondegenerate orbitals ϕ_A and ϕ_B, leading to the formation of two MOs: ψ_{MO} is the bonding combination obtained from the partial in-phase mixing of ϕ_B into ϕ_A and ψ_{MO}^* is the antibonding combination obtained from the partial out-of-phase mixing of ϕ_A into ϕ_B.

are now of different energies, the relative contributions of the AOs to the MOs are also different. The low-energy MO, ψ_{MO}, is obtained through the *in-phase* mixing of a small contribution of ϕ_B into ϕ_A, while the high-energy MO, ψ^*_{MO}, is obtained through the *out-of-phase* mixing of a small contribution of ϕ_A into ϕ_B. In other words the *low*-energy MO is made up predominantly from the *low*-energy AO, while the *high*-energy MO is made up predominantly from the *high*-energy AO. This mixing process may be expressed mathematically as

$$\psi_{MO} = N(\phi_A + \lambda\phi_B) \tag{2.49}$$

$$\psi^*_{MO} = N(\phi_B - \lambda\phi_A) \tag{2.50}$$

where λ is a mixing parameter whose value is less than 1, and N is the normalization constant, whose magnitude will depend on λ. The pictorial representation of the MOs (Fig. 2.9) reflects this unequal mixing by the different sizes of the two AOs in the MOs.

From perturbation theory it can be shown that the magnitudes of the mixing parameter λ, and the stabilization energy *SE*, are *inversely* proportional to the energy difference between the orbitals ΔE. Thus,

$$\lambda = \beta/\Delta E \tag{2.51}$$

$$SE = \beta^2/\Delta E \tag{2.52}$$

In other words, the *smaller* the energy gap, the *larger* both the mixing and the stabilization energy become. (Note that for small ΔE the perturbation equations no longer apply. As perturbation equations, they are derived on the assumption that ΔE is large.)

As in the case of the degenerate interaction, the actual energetic effect of the nondegenerate interaction depends on the electron occupancy of the two interacting orbitals. The most common situation for nondegenerate orbital interaction is the two-electron stabilization interaction, shown in 2.2. In this case an occupied low-energy orbital ϕ_A interacts with an unoccupied high-energy orbital ϕ_B in a stabilizing fashion. In physical terms the stabilization arises from the partial *delocalization* of the electronic charge into ϕ_B. In other words the two electrons are no longer localized within ϕ_A but located in an MO comprised of a linear combination of ϕ_A and ϕ_B, as illustrated in Fig. 2.9.

The significance of the HOMO and LUMO now becomes apparent since the interaction between nondegenerate orbitals is inversely proportional to the energy gap between the orbitals. Indeed the dominant role these orbitals often play has led to the development of frontier molecular orbital (FMO) theory, where chemical arguments are made based on the energy and symmetry properties of just these orbitals. The importance of the frontier orbitals can be demonstrated in a general fashion by considering the approach of two molecules A and B, with the MO orbital occupancies shown in **2.3.** In principle, two-electron stabilizing interactions are generated between *any* occupied orbital of

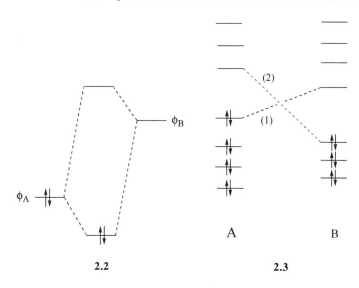

2.2   2.3

A that interacts with *any* unoccupied orbital in B, and vice versa. However, due to the role of the energy gap in determining the magnitude of the stabilization energy, of all the possible two-electron interactions, the two HOMO–LUMO interactions, labeled 1 and 2 in **2.3,** are likely to be most important, since it is for these two interactions that the energy gap is least. In fact, for the situation depicted in **2.3,** it would appear that interaction 1 will be the energetically dominant one, since the HOMO–LUMO energy gap for interaction 1 is smaller than for interaction 2. Note that when interaction 1 is dominant, the result is that molecule A acts as an electron donor toward molecule B, leading to partial electronic charge transfer from A to B.

Of course from Eq. (2.52) we see that the stabilization energy is a function of the resonance integral β, and not just the energy gap ΔE, so that the form of the orbital and its symmetry will also play a role in determining the stabilization energy. For this reason it is not always the HOMO–LUMO interaction that dominates structure and reactivity. Overlap effects, which are incorporated into β, may override energy gap considerations so that the qualitative analysis of orbital interactions is not always straightforward. However, despite these obvious limitations, FMO theory is widely applicable and has been used to successfully rationalize an enormous range of chemical phenomena. Some areas of chemistry where FMO arguments are commonly considered include molecular structure and conformation, reactivity and reaction stereochemistry, allowed and forbidden pericyclic reactions, solvation, charge-transfer complexes, to mention a few.

2.2.5 Rules for Determining Relative Orbital Energies

Given that we want to be able to interact orbitals with one another, it becomes important to have some guide to the *relative energies* of different orbitals.

Qualitative MO considerations often rely on little more than knowing the proper energetic ordering of interacting orbitals. Some useful guidelines for ordering orbital energies are

1. Bonding < Nonbonding < Antibonding

From Koopmans' theorem (Section 2.2.1) we know that ionization energies are a useful measure of occupied orbital energies (generally the HOMO), while electron affinities provide us with unoccupied orbital energies (generally the LUMO). For example, the first ionization potential of methane (12.5 eV) measures the HOMO energy level for that molecule. It is *greater* than the corresponding figure for methylamine (9.7 eV) because the HOMO in methane is *bonding* in character, while that in methylamine is a *nonbonding* nitrogen lone-pair orbital. Antibonding orbital energies are even higher in energy and may be estimated from the *electron affinity* of a molecule. (Note that to first approximation the electron affinity of a species is just the negative of the ionization potential of the anion of that species.) The fact that the electron affinities of most molecules are close to zero signifies that the LUMO level is high in energy compared with bonding and nonbonding MOs. Even for a molecule with a high electron affinity (i.e., which possesses a relatively *low*-energy antibonding orbital), such as tetracyanoethylene, the electron affinity is just 2.3 eV. This number, which needs to be compared with typical ionization potentials of neutral molecules that are in the 8–12 eV range, reflects the high energy of antibonding orbitals in general. In fact, it is worth noting that the LUMO energies of some molecules are *positive*. This means that the particular molecule has a *negative* electron affinity, that is, that the radical ion is unstable with respect to dissociation into the neutral molecule and a free electron. Not surprisingly, such radical anions are short lived.

2. σ Orbitals < π Orbitals < π* Orbitals < σ* Orbitals

Due to the better overlap between hybrid orbitals in the σ frame, σ orbitals are normally of lower energy than π orbitals. For the same reason π^* orbitals are normally lower in energy than σ^* orbitals. Thus it is generally easier to remove an electron from the π system of a molecule than from the σ system. Similarly, adding an electron to a molecule will normally result in the electron locating itself in a π^* orbital rather than a σ^* orbital. It is for this reason that a molecule commonly reacts through its π system rather than through its σ system; the π system normally constitutes the FMOs of the system and is therefore likely to be more reactive both to nucleophilic *and* electrophilic processes.

3. AO Energies Decrease Along a Row of the Periodic Table. Thus: C 2p > N 2p > O 2p > F 2p

The AO energy levels decrease along a row of the periodic table because of the increasing nuclear charge. For example, the first vertical ionization potential of CH_3NH_2 (9.7 eV) is less than that of CH_3OH (11.0 eV), reflecting the higher energy of the N lone pair in comparison to the O lone pair.

4. $\sigma_{C-C} > \sigma_{C-N} > \sigma_{C-O} > \sigma_{C-F}$

The C—X σ bond becomes a progressively *weaker* electron donor (i.e., it has a lower orbital energy) as one proceeds along the series from the C—C bond through to the C—F bond. This trend is a direct consequence of Rule 3 and may be readily demonstrated with a series of orbital interaction diagrams. The energy trend is illustrated schematically in Fig. 2.10a.

5. $\sigma^*_{C-C} > \sigma^*_{C-N} > \sigma^*_{C-O} > \sigma^*_{C-F}$

The C—X σ^* bond becomes a progressively *better* electron acceptor (i.e., it has a lower energy) as one proceeds along the series from the C—C antibonding orbital through to the C—F antibonding orbital. This trend is illustrated in Fig. 2.10a and is also a consequence of Rule 3.

6. $\sigma_{C-I} > \sigma_{C-Br} > \sigma_{C-Cl} > \sigma_{C-F}$

The valence shell energy levels of the four halogen atoms are quite similar, as reflected in their similar electron affinities, about 3.1–3.6 eV. For this reason,

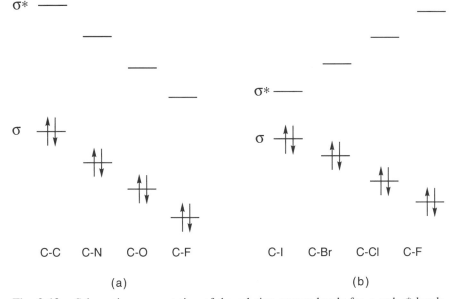

(a)   (b)

Fig. 2.10. Schematic representation of the relative energy levels for σ and σ^* levels for (a) C—C, C—N, C—O, and C—F bonds (i.e., along a row of the periodic table) and (b) C—F, C—Cl, C—Br, and C—I bonds (i.e., down a column of the period table).

on going down a column of the periodic table, it is C—X overlap that dominates the σ orbital energy. Since overlap between C and X decreases drastically from F to I, σ energy levels decrease from I to F.

7. $\sigma^*_{C-I} < \sigma^*_{C-Br} < \sigma^*_{C-Cl} < \sigma^*_{C-F}$

Overlap considerations are also responsible for the trend in this orbital series. Thus the σ orbital from Rule 6 that is the *least bonding* is paired up with the σ^* level that is the *least antibonding*. A diagram that illustrates the relative σ and σ^* energy levels for the C—X bonds is shown in Fig. 2.10b. Inspection of Fig. 2.10b makes it evident why organic fluorides are relatively unreactive to both oxidation and reduction, whereas organic iodides are reactive to both. The special stability of Teflon, a fluorocarbon polymer, can in part be understood in these terms.

2.2.6 Building Up MOs of Simple Molecules

A number of simple rules that are helpful in building up MOs in a qualitative way are now summarized.

Rule 1. The number of MOs generated from a basis set of AOs (or MOs) will equal the number of AOs (or MOs) in the set.

Rule 2. The mixing of two degenerate AOs (or MOs) leads to the formation of a low-energy *bonding* combination and a high-energy *antibonding* combination (Fig. 2.7).

Rule 3. When mixing two nondegenerate AOs (or MOs), two MOs are obtained (Fig. 2.9). The low-energy MO is obtained through the mixing of the high-energy AO (or MO) into the low-energy AO (or MO) in a *bonding* fashion, while the high-energy MO is obtained through the mixing of the low-energy AO (or MO) into the high-energy AO (or MO) in an *antibonding* fashion.

Rule 4. When a set of interacting orbitals can be mixed in a number of different orders, degenerate orbitals should be mixed together first so as to optimize symmetry.

Let us now illustrate these principles by going through some examples.

2.2.6A Ethylene. The π system of ethylene is built up from the interaction of the two C $2p$ orbitals. Interaction of the two AOs leads to the formation of two MOs. This is illustrated in Fig. 2.11 and the diagram is analogous to the one for σ-bond formation in the H_2 molecule. Thus the low-energy π orbital is obtained through the *in-phase* interaction of the $2p$ orbitals (i.e., mixing in a bonding fashion) while the high-energy π^* orbital is obtained through the out-of-phase interaction of these orbitals (i.e., mixing in an antibonding fashion). The two electrons that occupied the two AOs are allocated to the bonding orbital, as shown in Fig. 2.11.

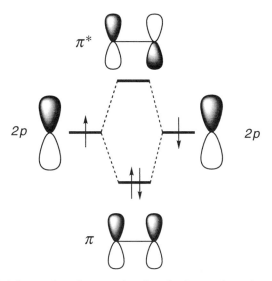

Fig. 2.11. Orbital interaction diagram showing the interaction of two degenerate C $2p$ orbitals to form the low-energy π orbital (occupied by two electrons), and the high-energy π^* orbital (unoccupied) of the ethylene molecule.

The π and π^* orbitals of ethylene form a basic building block in the construction of more complicated conjugated systems. Let us demonstrate this for the MOs of the allyl system.

2.2.6B Allyl. The π system of allyl is built up from the interaction of three C $2p$ orbitals, one on each C atom. Since our basis set consists of three AOs, we know the result of the interaction process will be to generate three MOs (Rule 1), and these may be built up in a step-wise fashion. Interestingly, the problem can be approached in two different ways, and it is instructive to look at both of these ways. We will see that an *unconventional* approach that makes optimal use of symmetry is much simpler, and hence is preferable, to the more conventional procedure.

Procedure 1 is the conventional approach where symmetry properties are not utilized.

Step 1. We begin by interacting the $2p$ orbitals on C1 and C2 to generate a set of π and π^* orbitals (as illustrated in Fig. 2.11 for ethylene).

Step 2. We now need to interact the third $2p$ orbital (the one on C3) with the π system just built up on C1 and C2. This step is not straightforward since the $2p$ orbital on C3 can interact with *both* π and π^* orbitals. Ideally, when interacting orbitals together the procedure should be carried out so that only 1 : 1 interactions take place. In the present case, however, this is not possible. The interaction diagram is illustrated in Fig. 2.12.

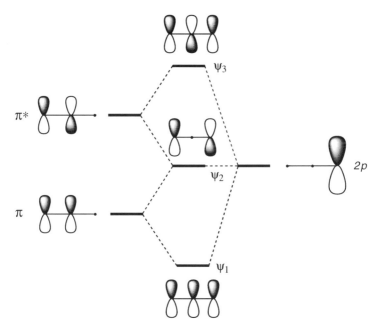

Fig. 2.12. Orbital interaction diagram showing the interaction of the π system of ethylene (on C1 and C2) with the C $2p$ orbital on C3 to generate the allyl MO system (*without* the aid of symmetry arguments).

Let us begin by considering the way the $2p$ orbital mixes into π and π^*, which is relatively straightforward. The $2p$ orbital mixes into the π orbital in a bonding fashion to generate the low-energy orbital ψ_1, while it mixes into the π^* orbital in an antibonding fashion to generate the high-energy orbital ψ_3. The form of ψ_2 is, however, more difficult to establish because it is obtained by the simultaneous mixing in of both π and π^* orbitals into the $2p$ orbital. The π orbital mixes into the $2p$ orbital in an *antibonding* fashion, since the π orbital is *lower* in energy than $2p$. This contribution is illustrated in **2.4** in Fig. 2.13. The π^* orbital mixes into the $2p$ orbital in a *bonding* fashion since the π^* orbital is *higher* in energy than $2p$ (illustrated by **2.5** in Fig. 2.13). These two interactions are not, however, *separate*. If they were, we would have generated *four* MOs from three AOs and this is not possible. So both of these contributions must now be merged into *one* overall interaction. This merger is done qualitatively by superimposing them on one another. This superimposition, illustrated in Fig. 2.13, generates ψ_2. Since the central orbitals on **2.4** and **2.5** have opposite phases, the result of the superimposition is the *cancelation* of the C2 coefficient; that is, the actual form of ψ_2 appears with a *zero* coefficient on C2.

It is apparent that this method of generating ψ_2 is quite cumbersome, not to mention imprecise (e.g., the complete cancelation of the C2 coefficient is not at all obvious). All the difficulties in this procedure arise from the fact that the

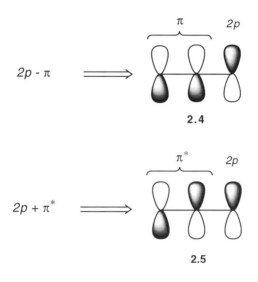

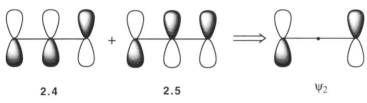

Fig. 2.13. Orbital interaction diagram showing the formation of ψ_2 in allyl based on the interaction of the C $2p$ orbital on C3 with π and π^* orbitals. Mixing of π into $2p$ in an antibonding fashion leads to orbital **2.4**, while mixing in of π^* into $2p$ in a bonding fashion leads to orbital **2.5**. Superposition of **2.4** and **2.5** generates ψ_2.

generation of ψ_2 comes about from a complex interaction of π and π^* with $2p$. Qualitative orbital mixing is most effective when it is done $1:1$, rather than $2:1$. Let us now look at an alternative method of building up the orbitals of allyl that is considerably simpler (and more precise), *relies on a powerful symmetry argument and ensures all interactions are 1:1.*

Procedure 2 is the approach where symmetry properties *are* utilized.

Step 1. We begin by formally interacting the $2p$ orbitals on C1 and C3 to generate an extended or long π bond. This may look odd at first sight since these two orbitals barely interact. However, we choose C1 and C3 (rather than C1 and C2) because these two carbon atoms *are equivalent by symmetry* and this will simplify the subsequent analysis. The set of π and π^* orbitals generated in this way, which we label $1 + 3$ and $1 - 3$, is illustrated in Fig. 2.14a. Note that in this case the π and π^* orbitals are not really bonding and antibonding orbitals. In fact, the two orbitals are *almost degenerate* due to the large C1–C3 distance, which decreases the overlap of the two p

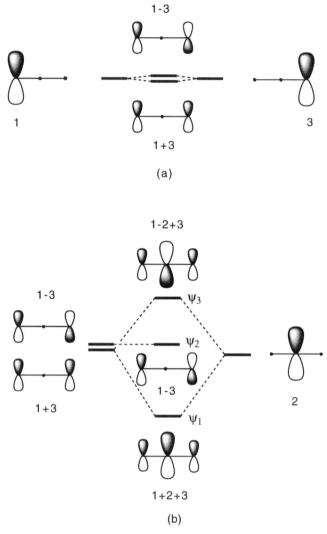

Fig. 2.14. Orbital interaction diagram showing the buildup of the π MOs of allyl (*with* the aid of symmetry arguments). (a) Initial degenerate mixing of C $2p$ orbitals on C1 and C3 to generate an effectively degenerate π and π^* pair $1 + 3$ and $1 - 3$. (b) Interaction of the π and π^* degenerate pair $1 + 3$ and $1 - 3$ with the C $2p$ orbital on C2 to generate the allyl MO system.

orbitals to almost zero (in contrast to the normal π and π^* orbital pair, illustrated in Fig. 2.11). *So in essence this step just converts two localized AOs into the delocalized MO representations of those AOs.*

Step 2. We now interact the C2 $2p$ orbital, which we label 2, with the $1 + 3$ and $1 - 3$ pair derived in Fig. 2.14a. This is illustrated in Fig. 2.14b. The crucial point in this interaction diagram is that $1 - 3$ *cannot interact with*

2 because these two orbitals are of different symmetry. The *2* orbital is *symmetric* with respect to a vertical mirror plane that bisects the C2 orbital while the *1 − 3* orbital is *antisymmetric* with respect to that plane. The result—this noninteracting orbital *1 − 3*, ends up as one of the final MOs of the allyl system (ψ_2). The other two orbitals ψ_1 and ψ_3 are obtained through the interaction of *1 + 3* with *2* in a bonding and an antibonding manner, respectively: The in-phase combination generates ψ_1, while the out-of-phase combination generates ψ_3 (Fig. 2.14b). Interestingly, even the relative magnitudes of the coefficients in ψ_1 and ψ_3 (larger on C2; smaller on C1 and C3) come out simply in this qualitative procedure. Since the coefficients on C1 and C3 in *1 + 3* and *1 − 3* are 0.707 [0.707 is just the value of $1/\sqrt{2}$ that is obtained from the normalized wave function; See Eqs. (2.28) and (2.29)], and that in *2* is equal to 1, the contribution of C2 to the MO, will be *larger* than the contribution of C1 and C3. Of course, this argument also relies on the near degeneracy of *2* and *1 + 3*.

The qualitative result obtained above conforms with the quantitative Hückel analysis for the allyl system (see Section 2.1.4). The lesson here is that utilizing symmetry in an optimal way can simplify the interaction diagram considerably. The main effect in the example discussed, is that all interactions become 1:1, thereby avoiding the complications that occur with 2:1 (or even more complicated) interactions, that do not lend themselves to a simple qualitative treatment.

2.2.6C Butadiene. The π system of butadiene is built up from the interaction of four C *2p* orbitals, one on each C atom. Since our basis set consists of four AOs we know the interaction process will generate four MOs. As in the case of the allyl system, there are a number of ways to build up this system. Conventionally, one might first interact C1 with C2, and C3 with C4 to generate two π systems, and then interact the two π systems that result with one another. This procedure, however, runs into the same difficulties we discovered in our treatment of the allyl system. The problem is that each of the π and π^* orbitals of one π system interact with each of the π and π^* orbitals of the other π system. A simplifying assumption that the π orbital of the C1–C2 system does not interact with the π^* orbital of the C3–C4 system (and vice versa) is usually made so that the analysis is more tractable. (The reader should try this approach as an exercise.) However, a better procedure is, once again, to exploit the power of symmetry, and to build up the π MOs for the four carbon chain by first interacting the *equivalent* carbon atoms, and *not* the carbon atoms that simply happen to be adjacent. In this case the equivalent carbon atoms are the C1–C4 pair and the C2–C3 pair. Let us now describe this procedure, which is illustrated in Fig. 2.15.

Step 1. We begin by interacting the equivalent C orbitals, C1 and C4, to generate a delocalized pair of degenerate π and π^* type orbitals (labeled *1 + 4* and *1 − 4*, respectively). This is illustrated in Fig. 2.15a. By analogy

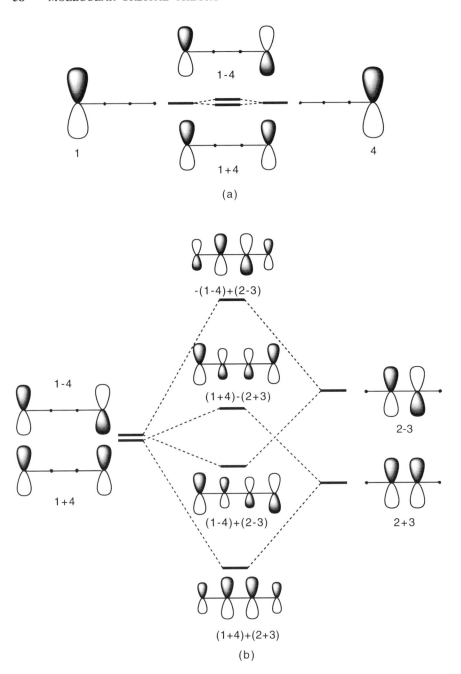

Fig. 2.15. Orbital interaction diagram showing the buildup of the π MOs of butadiene (*with* the aid of symmetry arguments). (a) Initial degenerate mixing of C $2p$ orbitals on C1 and C4 to generate an effectively degenerate π and π^* pair, $1 + 4$ and $1 - 4$. (b) Interaction of the degenerate pair, $1 + 4$ and $1 - 4$, with the $2 + 3$ and $2 - 3$ bonding and antibonding pair (π and π^* orbitals), to generate the butadiene MO system.

with the C1–C3 interaction in allyl, these two orbitals are also effectively degenerate since the large C1–C4 distance means that any overlap is close to zero.

Step 2. We now interact C2 and C3, the second pair of equivalent C orbitals, to generate a true π and π^* orbital set (as in ethylene; see Fig. 2.11). The resulting two orbitals are labeled $2 + 3$ and $2 - 3$, respectively.

Step 3. The two sets of π and π^* orbitals (the C1–C4 set and C2–C3 set) are now allowed to interact with each other. This process is illustrated in Fig. 2.15b.

What initially might appear to be a complex interaction diagram is actually much simplified because of symmetry. Orbital $1 + 4$ does not interact with $2 - 3$, and $1 - 4$ does not interact with $2 + 3$; in both cases there is a symmetry mismatch. Thus, by symmetry, $1 + 4$ can only interact with $2 + 3$ and $1 - 4$ can only interact with $2 - 3$. In other words by ordering the orbital interactions so as to optimize symmetry, only $1:1$ interactions take place. These are shown in Fig. 2.15. The resulting orbitals are

$$\psi_1 = (1 + 4) + (2 + 3)$$

$$\psi_2 = (1 - 4) + (2 - 3)$$

$$\psi_3 = (1 + 4) - (2 + 3)$$

$$\psi_4 = -(1 - 4) + (2 - 3)$$

In addition to providing us with the orbital energies and their phases, Fig. 2.15 also provides us with the *relative sizes of the AO coefficients in each orbital.* Since both interactions in Fig. 2.15b are between *nondegenerate* orbitals their *weight* in the final MOs are different. Thus ψ_1 will have a *larger* weight of $2 + 3$ than $1 + 4$, which means that the AO coefficients on C2 and C3 will be *larger* than on C1 and C4. This result can be confirmed from the Hückel analysis from which the coefficients are obtained quantitatively (see Section 2.1.4). A similar argument predicts the relative AO coefficients for the remaining MOs.

2.2.6D   Methane. The previous examples have all involved the building up of π MOs from a set of C $2p$ orbitals. However, there is no reason that σ-bonded systems cannot be described in this way as well. A simple molecule that illustrates the way the MOs of a σ system can be derived is the methane molecule (**2.6**). Our basis set of orbitals will include the four H $1s$ AOs and the C $2s$ and $2p$ AOs.

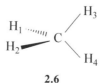

2.6

We begin to build up the MOs of methane (in its normal tetrahedral geometry) by interacting degenerate orbitals (Rule 1). First, let us start with the interaction of H1 and H2, and then follow that up with the interaction of H3 and H4 (numbering shown in **2.6**). In building up the hydrogen MOs, we initially ignore the C atom, leaving interactions of the C AOs to last.

The H1 and H2 AOs interact to generate a σ and σ^* MO system (just as in the hydrogen molecule). The resulting MOs, labeled $1 + 2$ and $1 - 2$, respectively, are illustrated on the left side of Fig. 2.16. H3 and H4 interact in exactly the same way to generate the $3 + 4$ and $3 - 4$ pair (illustrated on the right side of Fig. 2.16).

The next step is to interact these two sets of MOs, thus generating the MOs of the H_4 system. This is shown in Fig. 2.16. Once again symmetry greatly simplifies the interaction diagram. The $1 - 2$ MO does not interact with either $3 + 4$ or $3 - 4$, so it just becomes an MO of the H_4 system. The same holds for the $3 - 4$ orbital; it cannot interact with either $1 + 2$ or $1 - 2$. So the

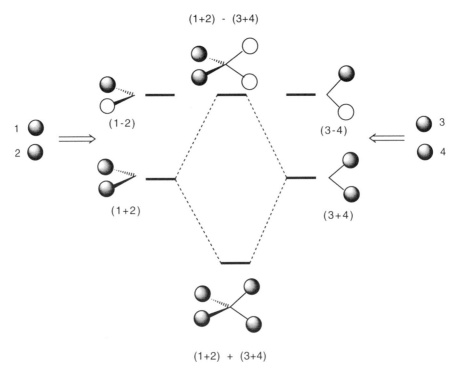

Fig. 2.16. Orbital interaction diagram showing the buildup of the set of four MOs for the tetrahedral H_4 system. Interaction is done pair wise; initially H1 and H2 are interacted together to give the $1 + 2$ and $1 - 2$ pair, and H3 and H4 to give the $3 + 4$ and $3 - 4$ pair. These two pairs are then interacted to give the H_4 MO set comprising a low-energy orbital, $[(1 + 2) + (3 + 4)]$, and three degenerate higher energy orbitals, $1 - 2$, $3 - 4$, and $[(1 + 2)-(3 + 4)]$.

only two orbitals that *are* allowed to interact by symmetry are *1 + 2* and *3 + 4*. This interaction (illustrated in Fig. 2.16) generates two MOs: a bonding combination, *(1 + 2) + (3 + 4)*, and an antibonding combination, *(1 + 2) − (3 + 4)*. Note that the antibonding combination, *(1 + 2) − (3 + 4)*, is degenerate with *1 − 2* and *3 − 4*, since all three MOs possess just one nodal plane. Thus the MOs of the tetrahedral H_4 system are a single low-energy bonding MO and a set of three degenerate higher energy MOs.[b]

The final step in generating the MOs of methane involves placing the C atom into the center of the H_4 tetrahedron to allow the interaction of the H_4 MO set with the set of *2s* and *2p* AOs on C (illustrated in **2.7**).

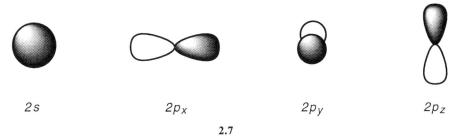

| 2s | 2px | 2py | 2pz |

2.7

At first sight one might think that attempting to interact four orbitals with four other orbitals is beyond the bounds of simple qualitative reasoning. However, closer inspection again reveals that symmetry greatly simplifies the problem. In every case there is only *one* H_4 orbital that can interact with any particular C AO. *All other interactions are symmetry forbidden* (the reader is encouraged to verify this statement). Thus,

> *(1 + 2) + (3 + 4)* can only interact with C *2s*
> *(1 + 2) − (3 + 4)* can only interact with C *2p_x*
> *1 − 2* can only interact with C *2p_y*
> *3 − 4* can only interact with C *2p_z*

Accordingly, interaction of the four H_4 orbitals with the four C *2p* AOs generates eight MOs: four bonding MOs and four antibonding MOs. These orbitals

[b]Inspection of the nodal charcteristics of the MOs makes it clear why tetrahedral H_4 would be an unstable species. Populating the MOs with the four electrons in this species would place two electrons into *(1 + 2) + (3 + 4)*, which is bonding between all four H atoms. However, the remaining two electrons would need to populate an antibonding orbital. If we were to place these two electrons into *(1 + 2) − (3 + 4)* then a distortion, in which H1 and H2 draw together (due to their bonding interaction), and pull away from H3 and H4 (with which they share an antibonding interaction), would lead to a lowering of the energy of this MO. This distortion would have a minimal effect on the energy of *(1 + 2) + (3 + 4)* since some bonding interactions (those between H1 and H2, and between H3 and H4) would *increase*, while others (those between H2 and H3 and between H1 and H4) would *decrease*. To summarize: *tetrahedral H_4 would want to distort away from the tetrahedral geometry and break up into two discrete H_2 units.*

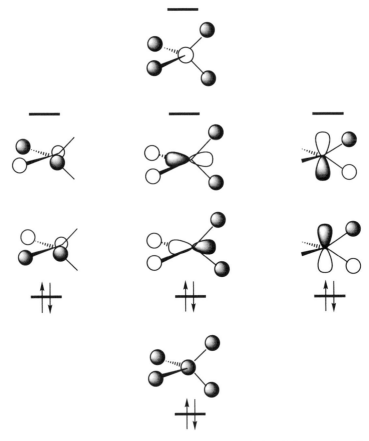

Fig. 2.17. Set of eight MOs (four bonding and occupied and four antibonding and unoccupied) for the CH_4 molecule.

are illustrated in Fig. 2.17. Given that eight valence electrons need to be placed in these orbitals, the four low-energy (and primarily bonding) MOs are occupied.

Several points need to be noted. First, whereas the source of the covalent bond between the two hydrogen atoms in H_2 is very clear from MO theory, for more complex molecules, the existence of a bond between two given atoms is less obvious. In building up the MOs of methane, we discovered that the MOs are spread over the entire system of atoms that make up the molecule. Consequently, the electrons that occupy those MOs are also delocalized over the molecule as a whole. Given that each MO can hold a maximum of two electrons, this means that each of the occupied MOs within the molecule often contributes just fractionally to the bonding interaction between any two atoms. In other words, according to MO theory the two-electron covalent bond between any two atoms, so readily represented by a simple Lewis structure, is the sum of many fractional bonding contributions, provided by each of the MOs. The *delocalized* nature of bonding, as described by MO theory, is in direct contrast to the *localized* nature of bonding, as described by VB theory. We will see in subsequent chapters that each approach has its particular advantages.

Second, the concept of hybridization, so central to the VB description of tetrahedral methane, is circumvented in the MO treatment. The MO description of methane simply does not require the prior mixing of $2s$ and $2p$ carbon AOs in order to generate the four sp^3 hybrids. [However, as we will see when we discuss the MOs of the methyl group (Section 2.2.6E), the concept of hybridization *will* be utilized in more complex MO mixing diagrams. It is only for some systems, such as methane, where symmetry greatly simplifies the orbital-mixing process, that hybridization does not need to be invoked].

The question now is do each of the C—H bonds in methane possess sp^3 hybrid character according to an MO description? Happily, the answer is yes. If we look at a particular C—H bond in methane and examine the nature of the bonding in that particular bond, the hybrid character of the bond is apparent within the MO framework as well. In methane three of the four occupied MOs contribute to each of the C—H bonds, so all three of these MOs determine the character (and strength) of that bond. It is the lowest energy MO that contributes the s character to the C—H bonds, since the carbon atom contributes its $2s$ orbital to that MO. Similarly, it is the higher energy bonding MOs that contribute the $2p$ character to those bonds. In other words the hybrid character of each C—H bond *is* present within the MO formalism, but it comes about through the combined contribution of all the bonding MOs.

A final point of interest concerns the photoelectron spectrum of methane. According to a simple VB picture, it would be concluded that all eight valence electrons of methane are equivalent, and therefore isoenergetic. Actually, the valence electrons of methane display *two* peaks in the photoelectron spectrum (one at 13 eV and one at 27 eV) signifying that all eight valence electrons do not possess the same energy. This picture corresponds best with the delocalized MO description of methane, which places the valence electrons in MOs of different energies, than with the localized VB picture.

Does this mean the MO method is superior to the VB method? Not at all! As we pointed out previously, some aspects of reactivity may be more readily understood in terms of one method, while others may be more readily understood by the other. Solving chemical problems and skinning cats have at least one feature in common!

2.2.6E   The Methyl Group. Discussing the MOs of methane has didactic value, but is in itself, of limited chemical use. More interesting from a chemical point of view is the methyl group. The methyl group is ubiquitous in chemistry and often has a profound effect on the chemistry of the substrate to which it is bound. For example, methyl chloride and *tert*-butyl chloride, related by three methyl groups, differ dramatically both in their solvolytic reactivity and the reaction mechanism that is followed. A proper understanding of the MOs of the methyl substituent is important if we want to understand how that group is likely to interact with substrates to which it is bound. Furthermore, the way the methyl group interacts with a substrate will provide insight into the way substituent effects operate in general.

The MOs of the methyl group may be built up by analogy with those of methane. The first step is to build up the MOs for the triangular H_3 system (where the three hydrogen atoms of a methyl group lie at the corners of an equilateral triangle). The second step is to interact this H_3 MO set with the carbon atom AOs. Let us now describe how this is done.

We begin by taking two H atoms and build up the σ and $\sigma*$ MOs of H_2 (this was illustrated in Fig. 2.7). To obtain the H_3 MOs we now need to interact these two MOs (which we label $1 + 2$ and $1 - 2$, respectively) with the third hydrogen 3. This is illustrated in Fig. 2.18. Again, we note that symmetry simplifies the interaction diagram. Orbitals $1 - 2$ and 3 are of different symmetry and so cannot interact. The only interaction that *is* allowed by symmetry is that of $1 + 2$ with 3, as shown in Fig. 2.18. This interaction yields a low-energy bonding combination $1 + 2 + 3$, and an antibonding combination $[-(1 + 2) + (3)]$ which, together with the noninteracting $1 - 2$ orbital, make up

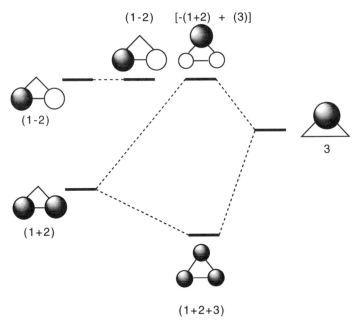

Fig. 2.18. Orbital interaction diagram showing the buildup of the set of three MOs for the triangular H_3 system. The orbital pair $1 + 2$ and $1 - 2$, obtained from the interaction of H1 and H2, are interacted with H3 to give the set of three MOs, comprised of the low-energy $1 + 2 + 3$ orbital, and the pair of degenerate orbitals $1 - 2$ and $[-(1 + 2) + (3)]$.

the H_3 set of three MOs. Note that $[-(1 + 2) + (3)]$ and $1 - 2$ are degenerate, though this may not be obvious from the qualitative treatment. The degeneracy can be demonstrated with a Hückel analysis, or rationalized qualitatively because both orbitals have one nodal plane.

We can now proceed to generate the methyl orbitals by interacting the set of three H_3 MOs with the set of carbon valence orbitals ($2s$, $2p_x$, $2p_y$ and $2p_z$), shown in **2.7**. Both sets of interacting orbitals are illustrated in Fig. 2.19. (The bond lines in the H_3 MOs are not part of the MOs but are drawn to clarify the three-dimensional geometry of the H_3 moiety, its orientation in space, and to indicate the position that the C atom will eventually take up within the methyl group. Note the different orientation of the H_3 moiety in Fig. 2.19, compared with that in Fig. 2.18, which will facilitate the CH_3 orbital description). Interaction of the above orbitals, though more complex than previous orbital interactions, may still be analyzed using qualitative considerations.

First, note that by symmetry, orbital $1 - 2$ can only interact with $2p_y$, while $[-(1 + 2) + (3)]$ can only interact with $2p_z$. However, with orbital $1 + 2 + 3$ there is a problem. By symmetry, this orbital can interact with both $2s$ and $2p_x$ orbitals and this complicates the qualitative argument (since, as we already mentioned, it is only 1:1 interactions that lend themselves to a simple qualitative treatment). The way around this difficulty is to mix together the $2s$ and $2p_x$ orbitals *prior* to interaction with the H_3 moiety. This mixing process leads to two *hybrid* orbitals—one pointing *toward* the H_3 group, *which will be involved in bonding with the hydrogen atoms*, and the other pointing *away* from that group, and which will *not* be involved in bonding. This mixing process is

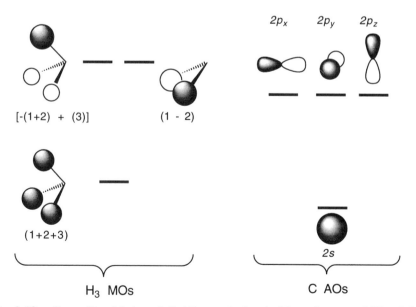

Fig. 2.19. Group H_3 orbitals and C AOs required to build up the group MOs of the methyl substituent.

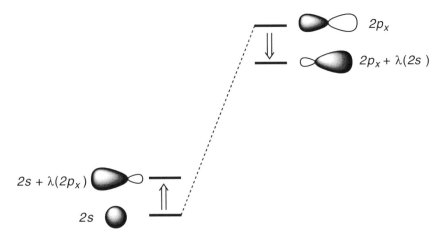

Fig. 2.20. Formation of two hybrid orbitals on C from the mixing of atomic $2s$ and $2p_x$ orbitals. The lower energy orbital [$2s + \lambda(2p_x)$] is primarily s in character, while the higher energy orbital [$2p_x + \lambda(2s)$] is primarily p in character. Note that this hybridization, or mixing process, is distinct from orbital interaction that cannot occur for orthogonal orbitals.

illustrated in Fig. 2.20. The hybrid orbital that points *toward* the hydrogen atoms (to the left) is largely $2s$ in character, and is the orbital that interacts with the $1 + 2 + 3$ H_3 group orbital, while the hybrid orbital pointing *away* from the hydrogen atoms (to the right), largely $2p$ in character, is the free sp^3 hybrid orbital (singly occupied) through which the methyl group will bond to some substrate.

Note that the prior mixing of C $2s$ and C $2p_x$ is *not* an orbital interaction of the kind we have been considering, since the two AOs are orthogonal, and therefore *cannot* interact. (By definition orthogonal orbitals have different symmetries and therefore cannot interact.) It is for this reason that the mixing process is just that—*a mixing process that generates two hybrid orbitals whose energies are a weighted average of the two original AOs* (Fig. 2.20). It is distinct from the usual interaction diagrams where stabilized bonding and de-stabilized antibonding combinations are obtained from the interaction of non-orthogonal orbitals. Thus we carry out this prior mixing of $2s$ and $2p_x$ orbitals for the sole purpose of facilitating subsequent interaction of *both* these AOs with the interacting orbital of appropriate symmetry (the $1 + 2 + 3$ H_3 group orbital). [We will subsequently use this argument to demonstrate that the oxygen atom in the water molecule is best thought of as sp^2 hybridized rather than sp^3 hybridized, as is commonly invoked (Section 2.2.8)].

The group orbitals of methyl can now be obtained by interacting the carbon AOs and hybrid orbitals with the H_3 group MOs. The set of MOs is illustrated in Fig. 2.21. Seven MOs are obtained (since the basis set comprises seven AOs): a low-lying occupied σ-type orbital labeled $\sigma(Me)$ (obtained from the

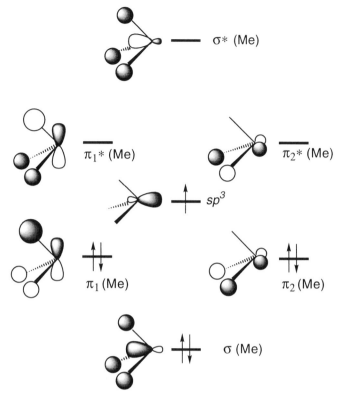

Fig. 2.21. Set of seven MOs for the methyl (Me) group comprised of three bonding and doubly occupied orbitals, σ(Me), π_1(Me), and π_2(Me); the singly occupied sp^3 hybrid, and three antibonding and unoccupied orbitals, π_1^*(Me), π_2^*(Me), and σ^*(Me).

in-phase mixing of $1 + 2 + 3$ with the C hybrid orbital), two degenerate occupied π-type orbitals labeled π_1(Me) and π_2(Me) (note that π_1(Me) is obtained from the in-phase mixing of $[-(1 + 2) + (3)]$ with $2p_z$, and π_2(Me) is obtained from the in-phase mixing of $1 - 2$ with $2p_y$), the singly occupied nonbonding hybrid orbital, and the set of three vacant antibonding orbitals, π_1^*(Me), π_2^*(Me), and σ^*(Me), obtained from the out-of-phase mixing of the corresponding orbitals.

The most striking aspect of this analysis is that the methyl group, ostensibly composed of just σ orbitals, *is perceived to contain four orbitals of π-type symmetry*: π_1(Me), π_2(Me), π_1^*(Me), and π_2^*(Me). Of course, these orbitals are not *actual* π orbitals since the three bonds between C and H are σ in character. However, the interaction of these σ bonds with each other effectively leads to the formation of group MOs *with overall π-type symmetry*. For this reason they may be termed *pseudo-π* orbitals. This point will be crucial in understanding the way a methyl group interacts with other π systems.

2.2.7 Conjugation and Hyperconjugation

Conjugation involves the interaction of one π system with another. The inter-action of two double bonds, or the interaction of a double bond with a benzene ring, are examples of *conjugative* interactions. The interaction of a *pseudo-π system* with a π *system* is termed *hyperconjugation*. Examples of hyperconju-gation are the interaction of a methyl group with a double bond (as in propene), or with a benzene ring (as in toluene). Interaction of two methyl groups (in which a pseudo-π system interacts with another pseudo-π system) may also be termed hyperconjugation. Some examples of conjugation and hyperconjugation are illustrated in **2.8**.

$$C=C\cdots C=C \qquad\qquad C=C\cdots CH_3$$
$$\pi \qquad\quad \pi \qquad\qquad\qquad \pi \quad\quad \text{pseudo-}\pi$$
$$\text{Conjugation} \qquad\qquad \text{Hyperconjugation}$$

$$C_6H_6\cdots CH_3 \qquad\qquad CH_3\cdots CH_3$$
$$\pi \quad\quad \text{pseudo-}\pi \qquad\quad \text{pseudo-}\pi \quad \text{pseudo-}\pi$$
$$\text{Hyperconjugation} \qquad\qquad \text{Hyperconjugation}$$

2.8

Now, let us consider the hyperconjugative interaction between a methyl group and vacant $2p$ orbital, as exemplified by the ethyl cation. The main orbital interaction that takes place between these two groups is between one of the $\pi(\text{Me})$ orbitals and the C $2p$ orbital. This is illustrated in Fig. 2.22. For the conformation illustrated (with the CH_2^+ segment perpendicular to the plane of the paper), the methyl MO of the appropriate symmetry that can interact with C $2p$ is $\pi_1(\text{Me})$. This two-electron stabilizing interaction between an occupied MO and an unoccupied MO has both energetic and structural con-sequences. First, the electron pair in $\pi_1(\text{Me})$ is stabilized by the mixing-in of a relatively low-lying vacant orbital. So this hyperconjugative interaction is *stabilizing*. Second, since the electron pair is now delocalized over the $H_3C—C$ moiety, and not just over the methyl group, various bond strengths and lengths will be affected. Specifically, the transfer of electron density to the region between the two carbon atoms will lead to $C—C$ bond *strengthening* (there is now some double-bond character in the bond) and, consequently, a *decrease* in the $C—C$ bond length (in comparison to a regular $C—C$ single bond). Of course this electron density has to come from somewhere and the source must be from the $C—H$ bonding region. Thus reduced electron density in the CH_3 region is predicted to lead to $C—H$ bond *weakening* within the methyl group and a corresponding *lengthening* of those $C—H$ bonds. Naturally, this inter-action also leads to the partial delocalization of the positive charge, since now there *is* some electron density in what was the vacant $2p$ orbital. All of these predictions are borne out by ab initio calculations on the ethyl cation.

If the conformation of the ethyl cation is changed so that the CH_2^+ moiety is in the plane of the paper, as shown in Fig. 2.23, then hyperconjugative

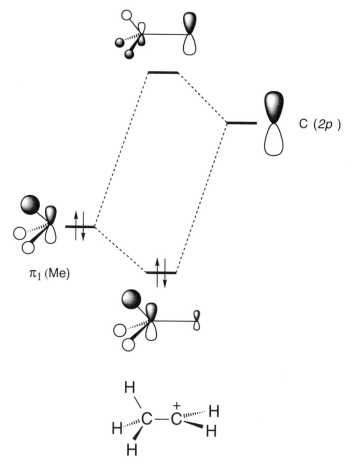

Fig. 2.22. Orbital interaction diagram showing the hyperconjugative stabilization in the ethyl cation in the perpendicular conformation. Two-electron stabilization occurs through the interaction of the occupied π_1(Me) orbital with the vacant C $2p$ orbital.

stabilization still results, but now symmetry requires that the orbital interaction that takes place is between the C $2p$ orbital and π_2(Me)—and not with π_1(Me). This is illustrated in Fig. 2.23. It is for this reason that the stability of the $CH_3CH_2^+$ cation is largely independent of its conformation about the C—C bond. Regardless of the $CH_3CH_2^+$ cation conformation, there is hyperconjugative stabilization of the cationic center from either (or both) π_1(Me) or π_2(Me).

Note that this MO description of hyperconjugation leads to essentially identical conclusions to a VB description of the phenomenon. The VB description, which is based on the mixing in of the so-called "no bond" resonance structures, such as **2.10** into the primary structure **2.9,** also suggests a shift of electron density from the methyl C—H bonds into the C—C bond, leading to

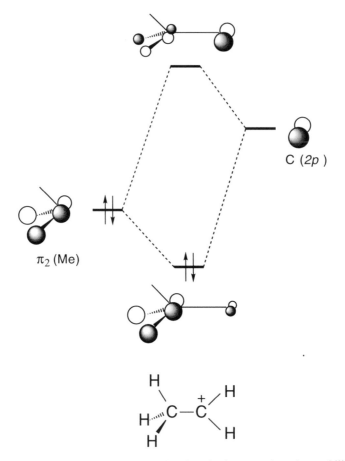

Fig. 2.23. Orbital interaction diagram showing the hyperconjugative stabilization of the ethyl cation in the eclipsed conformation. Two-electron stabilization occurs through the interaction of the occupied π_2(Me) orbital with the vacant C $2p$ orbital.

a weakening (and lengthening) of the C—H bonds together with a strengthening (shortening) of the C—C bond.

Of course, if the ethyl cation is substituted by some substituent X in the methyl group, then the stability of the two conformations, *perpendicular* **2.11** and *eclipsed* **2.12**, will differ; substitution lifts the degeneracy. For example, consider the case where X = F. Since the electronegative F atom contributes

2.9   **2.10**

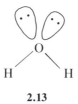

2.11   **2.12**

to the $\pi_1(CH_2F)$ orbital directly, the energy of $\pi_1(CH_2F)$ is lowered more than that of $\pi_2(CH_2F)$, where the F atom does not participate directly. As a result the stabilization brought about through the interaction of $\pi_1(CH_2F)$ with the C $2p$ orbital will be *less* than that with $\pi_2(CH_2F)$. Consequently, for the fluoro-ethyl cation, the eclipsed conformation, **2.12**, will be *more* stable than the perpendicular one, **2.11**.

2.2.8 Nonequivalence of Oxygen Lone Pairs in the Water Molecule

A picture of the water molecule that one commonly finds in chemistry texts shows an sp^3 hybridized O atom with two equivalent lone pairs on oxygen—the so-called "rabbit ears" picture **2.13**. This picture, while satisfactory for most purposes, is inferior to an alternative picture in which the two orbitals are *nonequivalent*. Let us examine this point.

2.13

The process of building up the MOs of water involves the interaction of the valence orbitals on oxygen ($2s$, $2p_x$, $2p_y$, and $2p_z$) with the two MO fragments associated with the two H atoms: the bonding $1 + 2$ and antibonding $1 - 2$ combinations. This is illustrated in Fig. 2.24a. If we check to see which of the orbitals on O are allowed by symmetry to interact with the H group orbitals, we find that the $2p_z$ orbital does not have the correct symmetry to interact with either $1 + 2$ or $1 - 2$. *This means that the oxygen $2p_z$ orbital remains unaffected and must be one of the final MOs in our MO construction of the water molecule.* The interactions that *are* allowed are between both the oxygen $2s$ and $2p_x$ orbitals with $1 + 2$, and between $2p_y$ and $1 - 2$. As discussed previously for the methyl group, the way of qualitatively dealing with the situation where two orbitals can interact with a particular orbital is to make a linear combination of those two orbitals (in this case $2s$ and $2p_x$) to form a hybrid orbital pointing toward the $1 + 2$ orbital, and a second hybrid orbital pointing away from the H atoms (that does not undergo interaction with the H group orbitals).

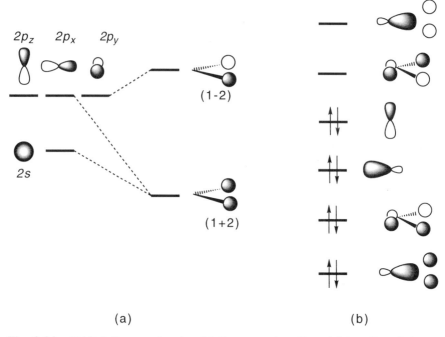

Fig. 2.24. Orbital diagram showing (a) the symmetry-allowed interactions between the set of O $2p$ orbitals and the two H_2 orbitals $(1 + 2)$ and $(1 - 2)$ in the H_2O molecule. (b) The set of MOs in the H_2O molecule obtained from the interaction of the orbitals in (a). Note that the two lone-pair orbitals in H_2O are *not* degenerate.

The full set of MOs of water that results from the interactions shown in Fig. 2.24a are illustrated in Fig. 2.24b. Thus we see that water possesses a total of four occupied MOs and two unoccupied MOs. The occupied MOs are two bonding MOs—a low-lying σ-type orbital and a higher energy orbital of π-type symmetry (due to a nodal plane), and two lone-pair orbitals—a lower energy sp^2 hybrid lone pair, and a "pure" $2p_z$ AO. The unoccupied MOs are two antibonding orbitals, one π type and one σ type. *Of special interest, however, is the fact that the two lone-pair orbitals, obtained in this interaction procedure, are not equivalent and therefore are predicted to have different energies.* Due to their different symmetry properties, one of the $2p$ orbitals (the $2p_z$ orbital) on O cannot undergo interaction with the H_2 group orbitals and remains unaffected.

At least one physical property of water is more consistent with this MO picture than the "rabbit ears" picture. The photoelectron spectrum of water shows that the two lone-pair orbitals on O are *nonequivalent and are separated by 2.1 eV.* So to the extent that the photoelectron spectrum can provide a picture of the MOs of a molecule, the MOs of Fig. 2.24 provide a better representation of the water molecule than the "rabbit ears" picture **2.13,** though it is important to appreciate that the two pictures are closely related, and may be transformed into one another by taking linear combinations of the two orbitals.

Applying the same kind of argument to the hydroxide ion (HO⁻) leads to the prediction that two of the three electron pairs on O in the hydroxide anion are "pure" $2p$ AOs, while the third one is an sp hybrid (along the HO bond axis).

The implications of this discussion to the question of hybridization can now be clarified. If one wishes to obtain MOs for a molecule that are consistent with the finding from photoelectron spectroscopy, the hybridization of the central atom (the premixing of its AOs to facilitate bonding) should only be carried out on those AOs of proper symmetry to interact with the ligand atom AOs. Thus the C atom in methane is sp^3 hybridized because the $2s$ and three $2p$ carbon AOs can all interact with the group MOs of the four H atoms. The C atom in the methyl cation is sp^2 hybridized because one of the $2p$ orbitals is no longer of the correct symmetry to interact with the group MOs of the three H atoms. On the same basis the O atom in water should be viewed as being sp^2 hybridized, while that in the hydroxide ion would be sp hybridized.

2.3 AROMATICITY

There is no precise definition for the term aromaticity[6a] or what constitutes an aromatic compound. The terms originally were associated with benzene or benzene-like derivatives, but nowadays the terms are used in a more general way to describe compounds that possess a closed cyclic system (normally of π symmetry, though systems of σ symmetry are also considered) that displays unusual stability—both kinetic and thermodynamic. Aromatic compounds are also characterized by certain magnetic properties due to the so-called *ring current*. The ring current, brought about by placing the aromatic molecule in an external magnetic field, may be detected by measuring proton nuclear magnetic resonance (^{1}H NMR) chemical shifts and magnetic susceptibilities.

A crude but simple measure of the special stability associated with an aromatic π system can be gauged from the heat of hydrogenation of the π system. For example, when benzene (C_6H_6) is hydrogenated to give cyclohexane (C_6H_{12})

$$\Delta H = -49.8 \text{ kcal mol}^{-1} \quad (2.53)$$

the heat of hydrogenation is surprisingly low. If we take the hydrogenation of cyclohexane as a reference for the heat of hydrogenation of a "normal" π bond

$$\Delta H = -28.6 \text{ kcal mol}^{-1} \quad (2.54)$$

we find that there is a substantial difference between the heat of hydrogenation of benzene (49.8 kcal mol^{-1}) and three times the heat of hydrogenation of cyclohexene, which is 85.8 kcal mol^{-1} (3×28.6 kcal mol^{-1}). The difference of 36 kcal mol^{-1} represents an approximate measure of the stability associated with the π system of benzene. This stabilization energy is commonly termed the *resonance energy* of the molecule.

Two points, however, need to be noted. First, the magnitude of the resonance energy can be estimated in different ways, so the value of 36 kcal mol^{-1} is not absolute, and should be treated as indicative. Second, the value of 36 kcal mol^{-1}, obtained from Eqs. (2.53) and (2.54), also includes "normal" resonance stabilization associated with any *acyclic* conjugated polyene. When we talk about aromatic stabilization we are speaking of the special stabilization of the conjugated *cyclic* system compared to a conjugated *acyclic* system. More recent approaches to the study of aromaticity attempt to obtain the "true" aromatic stabilization energy by subtracting this contribution from the overall resonance stabilization. These approaches lead to estimates of 20–25 kcal mol^{-1} for the aromatic stabilization energy of benzene.[6b]

In VB terms the stability of benzene has been explained by the resonance mixing of several valence forms, the two most important being

2.14

This simple description certainly explains the equal bond lengths of the C$-$C bonds (1.40 Å), which are intermediate between the standard C$-$C single- and double-bond lengths of 1.54 and 1.34 Å, respectively, but it does not explain the fact that the special aromatic character is also directly linked to the *number of π electrons in the ring*. A simple resonance picture would suggest that cyclobutadiene would also be stabilized, when in fact it is highly unstable. While a sufficiently rigorous VB treatment must also lead to a dependence between the number of electrons in the π system and the stability of the π system, an MO description of aromaticity is much simpler and directly relates the stability of the system to the number of π electrons.

2.3.1   Hückel *4n* + *2* Rule

In applying the principles of MO theory to cyclic π systems, (described in Section 2.1.4), Hückel made the observation that the *number* of electrons in the π system is expected to have a pronounced effect on the reactivity of the molecule: *Conjugated monocyclic π systems with 4n + 2 electrons are expected to show special stability, while those with just 4n electrons will be unusually reactive (n is an integer, 0, 1, 2, 3, · · ·).* The 4n + 2 π systems are termed *aromatic*, while the 4n systems are termed *antiaromatic*. The term antiaromatic signifies that *4n* systems are *destabilized* with respect to an acyclic conjugated system. Common aromatic and antiaromatic systems are shown in **2.15–2.20.**

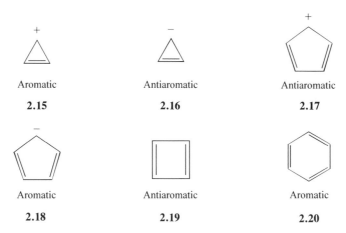

Aromatic	Antiaromatic	Antiaromatic
2.15	**2.16**	**2.17**
Aromatic	Antiaromatic	Aromatic
2.18	**2.19**	**2.20**

To understand the theoretical basis for expecting cyclic conjugated systems to be either stabilized or destabilized, we need to look at their MOs. These MOs can be found by solving the secular determinant as described in Section 2.1.4. However, a simple mnemonic that allows the π MO energies for any D_{nh} monocyclic system to be written down by inspection, rather than by solving the secular determinant, is available. The procedure is illustrated in Fig. 2.25.

One draws a circle of radius 2β into which the appropriate regular polygon is drawn with a vertex pointing down. This is demonstrated for three to six membered rings. The point of contact between each vertex and the circle represents the energy of an MO of the π system, with reference to a vertical energy scale. On this energy scale the center of the circle is the point of energy α, and the energies of the MOs are determined by specifying the vertical distance of the vertices from the center of the circle in units of β. In this way the energy of each MO, specified in terms of the parameters α and β, may be easily obtained without the need to set up and solve the Hückel determinant. The MO energy levels for the three to six membered rings obtained in this manner are shown in Fig. 2.25.

The orbital pattern obtained in this procedure can be generalized. In all cases there is a single low-energy orbital, and then above this orbital are located a set of degenerate orbital pairs. Finally, at the highest energy level one finds either a single MO if the polygon is even-numbered (cyclobutane or cyclohexane), or a degenerate pair of orbitals if the polygon is odd-numbered (cyclopropane or cyclopentane).

Inspection of Fig. 2.25 reveals that if electrons are added pairwise to the MO orbitals of cyclic conjugated systems, then the pattern of electron distribution will depend on whether the number of electrons placed in the system is *4n* or *4n + 2*. The addition of *4n + 2* electrons results in a set of fully occupied orbitals, while the addition of *4n* electrons results in a pair of two singly

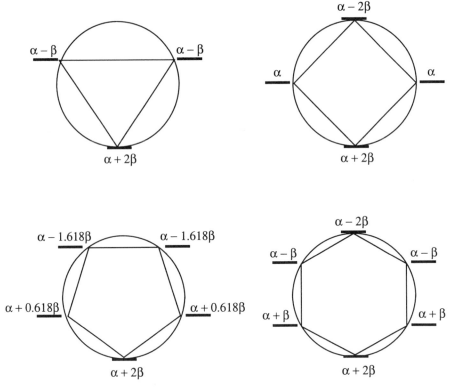

Fig. 2.25. Mnemonic for determining the π MO energies for a D_{nh} monocylic system, as illustrated for three to six membered rings (see text). The MO levels are determined by the point of contact of each vertex with the circle. The vertical axis represents an energy axis, where the center of the circle is of energy α and the radius of the circle is 2β. Corresponding MO energy levels expressed in terms of α and β are shown in the figure.

occupied orbitals: the last two electrons need to be assigned to a pair of degenerate orbitals. For example, the addition of either two or six electrons to the π system of a six-membered ring leads to a set of fully occupied orbitals, **2.21a** and **2.21c**, respectively, while adding either four or eight electrons to that system leads to the formation of two *singly* occupied orbitals, **2.21b** and **2.21d**, respectively. Since species with singly occupied MOs are likely to be more reactive than those with doubly occupied MOs, a simple rationalization of the $4n + 2$ rule is obtained.

Actually, the preceding argument is somewhat artificial because $4n$ systems do not need to have singly occupied orbitals. By undergoing a slight structural distortion, the degeneracy of the two HOMO orbitals can be lifted and the two electrons will then occupy the lower lying orbital that results. Consequently, the HOMO orbital in $4n$ systems may be doubly occupied. Indeed, ground-

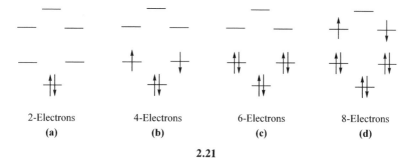

| 2-Electrons | 4-Electrons | 6-Electrons | 8-Electrons |
| (a) | (b) | (c) | (d) |

2.21

state cyclobutadiene is a singlet and rectangular in shape, rather than a triplet with a square geometry.

The greater stability of *4n + 2* species is probably better understood in terms of the energies of the occupied MOs, rather than by whether they are occupied by one or two electrons. Consider, for example, the cyclopropenyl system. From Fig. 2.25 we see that the set of three carbon $2p$ orbitals within a cyclopropenyl system generates three MOs: a low-lying MO of energy $\alpha + 2\beta$ and a higher lying pair of degenerate MOs of energy $\alpha - \beta$. Now let us check the effect of orbital occupancy on the energy of this system.

The placement of two electrons into this system generates the cyclopropenyl cation (**2.15**), while the placement of four electrons into the system generates the cyclopropenyl anion (**2.16**). Note, however, that the first two electrons go into a low-energy orbital ($E = \alpha + 2\beta$), while the next two electrons would go into a high-energy orbital ($E = \alpha - \beta$). *In fact, the second electron pair ends up at higher energy than the initial noninteracting electrons* (recall that the energy of a noninteracting $2p$ orbital is α). If we return to the cyclopropenyl cation, we see that the species has a low-energy HOMO and a high-energy LUMO. The low-energy HOMO will contribute to high thermodynamic stability, while the high-energy LUMO will lead to low reactivity. In the anion, however, the opposite behavior would be expected. In this case we have a species with a high-energy HOMO, which would lead to both low thermodynamic stability *and* high reactivity.

This energetic pattern, though less pronounced, is maintained for the other systems as well. Thus, for example, the four-membered ring system can accommodate just two electrons in a low-energy orbital ($E = \alpha + 2\beta$); a second electron pair (to generate neutral cyclobutadiene) would be accommodated in the relatively high-energy orbitals ($E = \alpha$). Hence, cyclobutadiene (**2.19**) is predicted to be unstable and indeed can only be isolated in a low-temperature matrix.

The five- and six-membered ring systems can accommodate six electrons in low-energy orbitals. Hence, the cyclopentadienyl anion (**2.18**) and benzene (**2.20**) are stabilized, but the cyclopentadienyl cation (**2.17**), with just four electrons and a low-lying LUMO orbital ($E = \alpha + 0.62\beta$), would be highly reactive.

2.3.2   Perturbation Theory Approach

A particularly intuitive way of understanding aromaticity, based on perturbation theory,[7] is to build up the cyclic π system from fragments. If the π interactions between these fragments are favorable the system is stable and aromatic. If the interaction is unfavorable the system is unstable and antiaromatic. This approach is readily illustrated with the cyclopropenyl cation and anion (**2.15** and **2.16**), and rests entirely on a symmetry argument.

The cyclopropenyl cation and anion systems may be generated from the interaction of the π system of ethylene and a lone $2p$ orbital. In the case of the cation an electron count tells us that this $2p$ orbital needs to be vacant, while for the anion the orbital must be occupied. An orbital diagram that considers the two situations is shown in Fig. 2.26.

In the case of the cation, interaction between the π system and the vacant $2p$ orbital is stabilizing and is allowed by symmetry. This is illustrated in Fig.

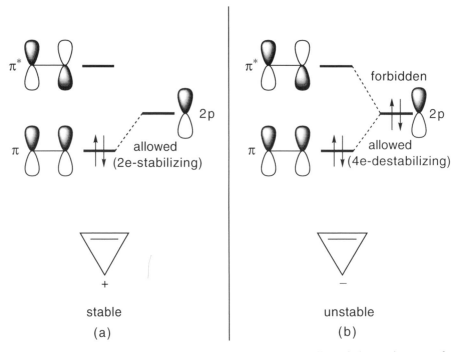

Fig. 2.26. Orbital interaction diagrams showing the stability of the cyclopropenyl cation and the instability of the cyclopropenyl anion. (a) Formation of the cyclopropenyl cation π system involves an *allowed* two-electron stabilizing interaction between an occupied π orbital and a vacant $2p$ orbital. (b) Formation of the cyclopropenyl anion π system from fragments involves an *allowed* four-electron destabilizing interaction between an occupied π orbital and an occupied $2p$ orbital. The potentially stabilizing two-electron interaction between the occupied $2p$ orbital and the vacant π^* orbital is symmetry forbidden.

2.26a. In the case of the anion, however, the potentially stabilizing two-electron interaction between the occupied *2p* orbital and the vacant π^* orbital *is forbidden by symmetry*. The interaction that *is* allowed by symmetry is now a four-electron *destabilizing* interaction between the occupied π and *2p* orbitals. This interaction is illustrated in Fig. 2.26b. Therefore the generation of the cyclopropenyl cation π system from π fragments is *stabilizing*, while the same procedure for the anion is *destabilizing*.

The instability of the cyclopentadienyl cation and the stability of the cyclopentadienyl anion are also readily demonstrated in this manner. An orbital interaction diagram for these two systems is illustrated in Fig. 2.27.

The cyclopentadienyl cation and anion may be generated from the interaction of the π orbitals of a butadiene fragment and a single *2p* orbital: vacant for the cation, occupied for the anion. For the cation the primary source of stabilization would be between the HOMO orbital of the butadiene fragment and the vacant *2p* orbital. However, this interaction, illustrated in Fig. 2.27a, is forbidden by symmetry. In the case of the anion, the two-electron stabilizing interaction is between the occupied *2p* orbital and the butadiene LUMO. This interaction *is* allowed by symmetry, as shown in Fig. 2.27b. In other words, for the cyclopentadienyl system, charge delocalization is inhibited for the cation, but is facilitated for the anion—precisely the opposite behavior to that observed in the cyclopropenyl system.

Other systems may be analyzed in the same way. As an exercise the reader can demonstrate that the cyclobutadiene system, generated from the interaction of two ethylene systems, leads to symmetry-allowed repulsive interactions and symmetry-forbidden stabilizing interactions, so the π interaction between these two molecules is *repulsive*. Significantly, this symmetry mismatch does not depend on the fragments that are used to build up the cyclobutadiene molecule, and may be demonstrated in different ways. Thus building up the cyclobutadiene π system from the coupling of an allyl fragment and a lone CH_2 fragment, leads to the same conclusion; there is a symmetry mismatch between the singly occupied allyl ψ_2 orbital and the singly occupied C *2p* orbital, so bonding is precluded.

On the other hand, building up benzene from butadiene and ethylene leads to stabilizing π interactions that *are* symmetry allowed, while the repulsive interactions are now symmetry forbidden, so the interaction between these two π systems is *favorable*. The same conclusion may also be reached by building up benzene from two allyl fragments (or a pentadienyl fragment with a lone *2p* orbital). The conclusion depends on the electron count and *not* the way the molecule is dissected.

Finally, it is worth pointing out that the *4n + 2* rule is not absolute but is governed by orbital topology. Consider the Molecule **2.22**. This molecule possesses six π electrons and so appears to conform to the *4n + 2* rule. However, the orientation of these orbitals is such that potential stabilizing two-electron interactions are symmetry forbidden; each of the occupied π orbitals cannot interact with the adjacent π^* orbitals due to a symmetry mismatch (as

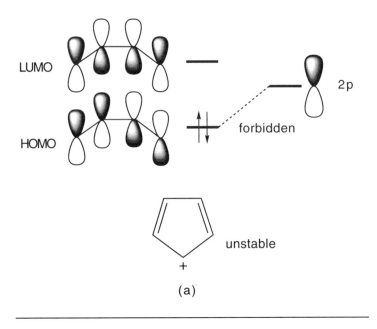

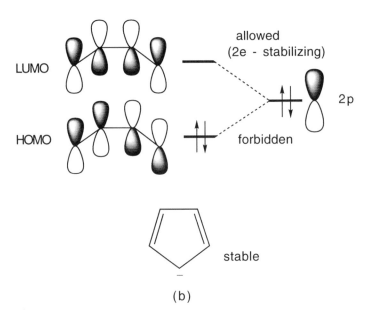

Fig. 2.27. Orbital interaction diagrams showing the instability of the cyclopentadienyl cation and the stability of the cyclopentadienyl anion. (a) Formation of the cyclopentadienyl cation π system from fragments involves a *forbidden* two-electron stabilizing interaction between the occupied butadiene HOMO and a vacant $2p$ orbital. (b) Formation of the cyclopentadienyl anion π system from fragments involves an *allowed* two-electron stabilizing interaction between the occupied $2p$ orbital and the vacant butadiene LUMO, while the four-electron *destabilizing* interaction between the butadiene HOMO and the occupied $2p$ orbital is symmetry *forbidden*.

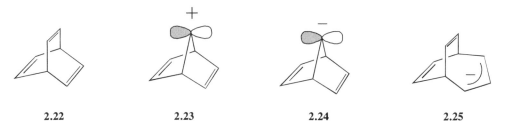

2.22   2.23   2.24   2.25

illustrated in Fig. 2.6, the overlap between π and π^* orbitals that approach head-on is zero by symmetry). However, the destabilizing four-electron $\pi-\pi$ interactions *are* symmetry allowed, leading to a destabilized molecule. Indeed, the relatively high heat of hydrogenation of **2.22** confirms the high energy of this six-electron π system.

Using similar perturbation arguments it is simple to show that the *4n* π system in the cation **2.23** is stabilized. The orientation of the two occupied π orbitals allows overlap with the vacant *2p* orbital so a stabilizing two-electron interaction takes place. On the other hand, the *4n + 2* π electron anion **2.24** *could not be prepared*, presumably because the occupied C *2p* orbital would not interact with the vacant π^* orbitals due to a symmetry mismatch. By contrast the eight π-electron anion **2.25** *is* stable due to the ability of the occupied high-lying allyl orbital (ψ_2 in Fig. 2.12) to interact with each of the vacant π^* orbitals in a stabilizing two-electron interaction.

Therefore, we see that the *4n + 2* rule is normally applied to planar monocyclic π systems; changing the orbital topology (such as in Möbius systems) can lead to an inversion in the rule, so that it is the *4n* systems that are the stable ones. Detailed rules that govern these topology changes have been developed.[8]

REFERENCES

1. For a comprehensive description of ab initio MO theory and its applications, see W. J. Hehre, L. Radom, P. v. R. Schleyer, and J. A. Pople, *Ab Initio Molecular Orbital Theory*, Wiley, New York, 1986. For an updated description, see J. Simons, *J. Phys. Chem.* **95,** 1017 (1991).

2. A derivation of the Variation Principle may be found in T. H. Lowry and K. S. Richardson, *Mechanism and Theory in Organic Chemistry*, 3rd ed., Harper & Row, New York, 1987, Chapter 1, Appendix 1.

3. For a recent account of various semiempirical methods, see M. J. S. Dewar, C. Jie, and J. Yu, *Tetrahedron*, **49,** 5003 (1993).

4. A description of the mathematical procedures that are needed to solve determinants and the set of LCAO coefficients may be found in Ref. 2.

5. For an excellent introduction to symmetry, see H. H. Jaffe and M. Orchin, *Symmetry in Chemistry*, Wiley, New York, 1965. A more advanced description may be found

in F. A. Cotton, *Chemical Applications of Group Theory*, 2nd ed., Wiley, New York, 1971.

6. (a) For a recent monograph on aromaticity, see V. I. Minkin, M. N. Glukhovtsev, and B. Ya. Simkin, *Aromaticity and Antiaromaticity. Electronic and Structural Effects*, Wiley, New York, 1994. (b) For different approaches to measuring the resonance energy in benzene, see M. J. S. Dewar and G. J. Gleicher, *J. Am. Chem. Soc.* **87,** 685, 692 (1965); H. Kollmar, *J. Am. Chem. Soc.* **101,** 4832 (1979); N. C. Baird, *J. Chem. Educ.* **48,** 509 (1971). For a VB approach to the problem, see S. S. Shaik, P. C. Hiberty, J-M. Lefour, and G. Ohanessian, *J. Am. Chem. Soc.* **109,** 363 (1987).

7. M. J. S. Dewar and R. C. Dougherty, *The PMO Theory of Organic Chemistry*, Plenum Press, New York, 1975.

8. M. J. Goldstein and R. Hoffmann, *J. Am. Chem. Soc.* **93,** 6193 (1971).

3

QUALITATIVE VALENCE BOND THEORY

We have already noted that in the limit MO and VB theories are theoretically equivalent, since both lead to the same state wave function when taken to a sufficiently high level of precision. However, as we noted in Chapter 1, VB theory has an important advantage over MO theory: VB theory is conceptually closer to the organic chemists' way of thinking about molecules. A pictorial representation of VB wave functions converts mathematical expressions into the well-known resonance structures of organic chemistry.

Modern organic chemistry still makes extensive use of the resonance idea, mainly to explain the structure of delocalized species, such as the enolate ion and benzene. However, the enormous developments in MO theory that took place in the 1960s and 1970s, both quantitative and qualitative, led to a perception that VB ideas are somewhat out-dated and inferior to the more modern MO approach. This view has arisen in large part because the explosive development in computational methods has almost entirely relied on the MO methodology. The reason for this choice, however, does not derive from any inherent theoretical correctness. Rather, the choice is governed by practical considerations, namely, which methodology is more readily converted into a computer algorithm. It turns out that the VB procedure is computationally less tractable than the MO one.

However, it will be stating the obvious if we say that the human brain and the electronic computer "think" very differently. As a result, in the scramble to embrace MO thinking possible advantages that the VB approach might provide have often been overlooked. So while it is true that certain chemical concepts (primarily those that depend on symmetry considerations) are inherently more suited to an MO description, it is also true that certain facets of structure and reactivity are more suited to a VB approach. In this and the following chapter we discuss some qualitative aspects of the VB methodology and explore their capabilities.

3.1 VALENCE BOND CONFIGURATION-MIXING DIAGRAMS

We have previously described the VB procedure for constructing electronic state wave functions from a basis set of AOs (Chapter 1). Two stages were involved: first the construction of the VB configurations, each of which provides a simple state description, and then the mixing of these configurations to generate an improved state description. *This second stage of the process, the mixing of VB configurations to generate improved state descriptions, is at the heart of VB theory* and forms the subject of much of this chapter. Note that this procedure is just the more formal quantum mechanical representation of *resonance mixing.* Given the importance of the resonance concept in chemistry, the central role of VB configuration-mixing (VBCM) becomes easier to appreciate.

If the state wave function is constructed from just two (or in some cases three) VB configurations, then the mixing process can be usefully represented in a *VBCM diagram.* Valence bond configuration-mixing diagrams are of great utility for understanding a variety of chemical phenomena; on the one hand they reflect basic quantum principles, yet, on the other hand, they lend themselves to a simple, nonmathematical treatment. In effect, these diagrams represent a middle road between the formal quantum mechanical application of VB theory and qualitative resonance theory. This chapter will demonstrate how VBCM diagrams may be used to understand a range of chemical concepts, including the nature of the covalent bond.

Let us begin with a description of the states of the $R-X$ bond, where R is some alkyl group and X is a halogen.

3.2 STATES OF THE $R-X$ BOND

The ground state of the $R-X$ bond, where $R-X$ represents a general alkyl halide, has both covalent and ionic character. Therefore it may be simply described in terms of a hybrid of covalent and ionic configurations. Thus, $\Psi_g(R-X)$, the ground-state wave function, may be written as

$$\Psi_g(R-X) = (R\cdot \ \cdot X) + \lambda(R^+ :X^-) \qquad (3.1)$$

which is equivalent to the organic chemist's representation

$$R-X = (R\cdot \ \cdot X) \leftrightarrow (R^+ :X^-) \qquad (3.2)$$

where the mixing-in of the ionic configuration imparts polar character to the covalent bond.

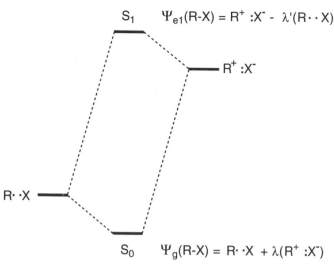

S_1 $\Psi_{e1}(R\text{-}X) = R^+ :X^- - \lambda'(R\cdot\cdot X)$

$R^+ :X^-$

$R\cdot\cdot X$

S_0 $\Psi_g(R\text{-}X) = R\cdot\cdot X + \lambda(R^+ :X^-)$

Fig. 3.1. Configuration-mixing diagram showing the mixing of the covalent configuration $R\cdot\cdot X$ and the ionic configuration $R^+ :X^-$, to generate improved descriptions of the ground (S_0) and first excited (S_1) states of the R—X bond. Note the ground state is primarily covalent in character, while the excited state is primarily ionic in character.

Now, let us describe the mixing process with the aid of a configuration-mixing diagram, as illustrated in Fig. 3.1. Such a diagram has similarities to an orbital interaction diagram, where two AOs interact together to generate two MOs. *In a configuration-mixing diagram, however, two configuration wave functions mix with one another to generate improved representations of two state wave functions.* Thus, according to Fig. 3.1 the high-energy ionic configuration $R^+ :X^-$ mixes into the low-energy covalent (Heitler–London) configuration $R\cdot\cdot X$ in a *stabilizing* way to generate the *ground-state* wave function of the bond $\Psi_g(R—X)$. Similarly, the low-energy covalent configuration mixes into the high-energy ionic configuration in a *destabilizing* way to generate the *first excited-state* wave function of the R—X bond $\Psi_{e1}(R—X)$

$$\Psi_{e1}(R—X) = (R^+ :X^-) - \lambda'(R\cdot\cdot X) \qquad (3.3)$$

The important point here is that the mixing of the *two* configurations leads to an improved description of *two* states—a ground state S_0 and an excited state S_1. Just as with orbital interaction, *in-phase* mixing leads to a description of the ground state, while *out-of-phase* mixing leads to a description of the excited state. This configuration-mixing diagram also shows that while the ground state of the bond is primarily *covalent* in character, the first excited state of the R—X bond is primarily *ionic* in character. In fact, to first approximation one can describe either state by the configuration that dominates the state description (i.e., $R\cdot\cdot X$ for the ground state and $R^+ :X^-$ for the first excited state).

Of course one could obtain a more accurate electronic description of the

R—X bond by mixing in a third configuration as well—the most reasonable possibility being R:$^-$ X$^+$. The interaction of *three* configurations leads to a more complex mixing diagram (shown in Fig. 3.2) in which now *three* state descriptions are generated—the ground state and *two* excited states. This simple picture predicts that the second excited state of the R—X bond will also be ionic in character—but with reverse polarity (i.e., it will be largely described by R:$^-$ X$^+$).

Note again, the analogy with the orbital-mixing procedure; just as the number of AOs that interact equals the number of MOs generated, so the number of configurations that are used in the mixing process equals the number of state descriptions that are generated. Also, the larger the number of configurations that are utilized, the more accurate our state descriptions become. A state description that is built up from three contributing configurations will be more precise (i.e., a better representation of the actual state) than one using just two configurations. Of course, additional configurations may improve the state picture even more (e.g., R^{2+} X^{2-}), but the high energy of these additional configurations means that their contribution to the state wave function (and hence their energetic impact) would be small.

One obvious difference between an orbital-mixing diagram and a configuration-mixing diagram is that in the configuration mixing diagram the configuration energy levels are never drawn with electrons occupying them. This

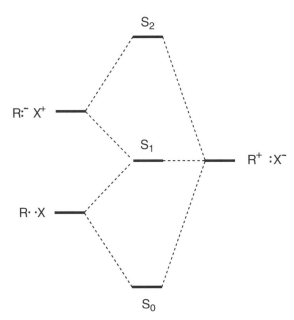

Fig. 3.2. Configuration-mixing diagram showing the mixing of the covalent configuration R· ·X and two ionic configurations R$^+$:X$^-$ and R:$^-$ X$^+$, to generate the ground (S_0) and first two excited states (S_1 and S_2) of the R—X bond.

occurs because an electronic configuration, by definition, has its electrons allocated to specified orbitals within that configuration, and the electrons are not therefore shown in the diagram. By contrast, in an orbital-mixing diagram, a particular orbital can be drawn as either vacant or occupied, so that the electron occupancy of a particular orbital is a central feature of an orbital-mixing diagram.

In constructing a configuration-mixing diagram, even in a schematic fashion, the relative energies of the interacting configurations need to be known. The two most important physical parameters that are needed to determine the relative energies of a set of configurations are the *ionization potential I* and *electron affinity A* of atoms and groups. This results because configurations are very often related by a single electron shift. Let us demonstrate this by returning to our example of the interaction of R· ·X with R$^+$:X$^-$.

Conversion of R· ·X into R$^+$:X$^-$ requires that an electron be transfered from R to X. Thus the energy gap ΔE, between these two configurations, when the two groups R and X are infinitely separated, is given by

$$\Delta E = I_R - A_X \tag{3.4}$$

where I_R is the ionization potential of R, and A_X is the electron affinity of X. For any two neutral groups at infinity, ΔE is always positive (i.e., the ionic configuration is higher in energy than the covalent configuration) since the *ionization potentials of neutral groups are always larger than the electron affinities of neutral groups.*

The actual magnitude of ΔE may be estimated from ionization potential and electron-affinity data (some of which are shown in Tables 7.1–7.5). The data confirm that *removing* an electron from a neutral species requires more energy than that released when an electron is *added* to a neutral species. From the data in the tables the value of ΔE for methyl chloride at infinite separation is calculated to be 143.5 kcal mol^{-1} (226.9 − 83.4).

The reason R:$^-$ X$^+$ is higher in energy than R$^+$:X$^-$ (Fig. 3.2) stems from the fact that both the ionization potential *and* electron affinity of X are higher than the corresponding values for R. Thus it requires *more* energy to transfer an electron from the more electronegative X to the less electronegative R than the reverse.

3.3 SINGLET AND TRIPLET STATES OF HYDROGEN

In our discussion of the states of the R—X bond we have considered the configuration R· ·X as a single configuration in the same way that we considered H· ·H to be the covalent configuration of the hydrogen molecule. Actually covalent configurations, such as R· ·X or H· ·H, are themselves, strictly speaking, a hybrid of two contributing configurations.

Consider two hydrogen atoms approaching one another to form a hydrogen molecule. Since the spins on these two H atoms can be either α or β, four possible spin combinations may result. At first sight these appear to be

$$H \cdot \uparrow \; \uparrow \cdot H \qquad \text{labeled as } \alpha\alpha$$

$$H \cdot \uparrow \; \downarrow \cdot H \qquad \text{labeled as } \alpha\beta$$

$$H \cdot \downarrow \; \uparrow \cdot H \qquad \text{labeled as } \beta\alpha$$

$$H \cdot \downarrow \; \downarrow \cdot H \qquad \text{labeled as } \beta\beta \qquad (3.5)$$

Actually, these four configurations do *not* represent the four spin states of the hydrogen molecule. While $\alpha\alpha$ and $\beta\beta$ *do* represent proper spin states, $\alpha\beta$ and $\beta\alpha$ do not, because these configuration spin functions violate a basic principle of quantum mechanics—electrons are indistinguishable. The $\alpha\beta$ function, for example, labels one electron as α and the other as β. However proper state functions for the $\alpha\beta$ and the $\beta\alpha$ pair that do not violate this basic principle may be obtained by taking linear combinations of these two configurations. This mixing process may also be carried out with a configuration-mixing diagram, as is shown in Fig. 3.3. We see that the two configurations $\alpha\beta$ and $\beta\alpha$ may combine in both positive and negative ways. As noted earlier, it is the negative combination of the two spin forms, $\alpha\beta - \beta\alpha$, that generates the Heitler–London *singlet* ground-state wave function [Eq. (1.5)]. Now we see that the *positive* combination $\alpha\beta + \beta\alpha$ generates one of the three triplet states of H_2

$$\Psi_{\text{Triplet}} = \frac{1}{\sqrt{2}} (\alpha\beta + \beta\alpha) \qquad (3.6)$$

(the other two triplet states are just the two more obvious forms $\alpha\alpha$ and $\beta\beta$).

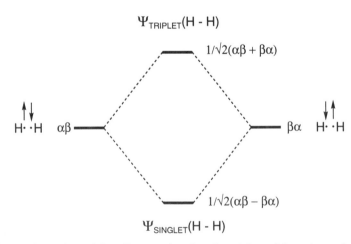

Fig. 3.3. Configuration-mixing diagram showing the mixing of the spin configurations $\alpha\beta$ and $\beta\alpha$, to generate two states of H_2; the singlet ground state described by $(1/\sqrt{2})(\alpha\beta - \beta\alpha)$, and the triplet excited state described by $(1/\sqrt{2})(\alpha\beta + \beta\alpha)$.

At first it might look rather strange that the triplet state may be represented by the wave function of Eq. (3.6), which is composed of spin-up and spin-down components. However, to the eternal sorrow of nontheoretical chemists, a precise quantum mechanical description of a system cannot always be readily simplified, and not all quantum phenomena can be reduced to a simple intuitive model. The third triplet state, depicted in Eq. (3.6), is one such example where a simple intuitive approach to electron spin is inadequate. In subsequent applications triplet states will be represented by the pictorial form $H \cdot \uparrow \uparrow \cdot H$ or $\alpha\alpha$, even though we can see that the wave function for the triplet state, as described in Eq. (3.6), is not properly represented by this symbolism.

The importance of the preceding discussion is that it demonstrates that from a VB perspective, configuration mixing is at the very heart of chemistry. *Even the covalent bond, a cornerstone of chemistry, may be viewed as an example of configuration mixing.* Restating this in the language of resonance, we may say that not just "special" molecules like benzene and the allyl radical are stabilized by resonance mixing; the stability of the covalent bond in the hydrogen molecule may also be attributed to resonance mixing as depicted by

$$H-H = (H \cdot \uparrow \downarrow \cdot H) \overset{(-)}{\longleftrightarrow} (H \cdot \downarrow \uparrow \cdot H) \tag{3.7}$$

3.4 RULES FOR MIXING OF VALENCE BOND CONFIGURATIONS

The mixing of configurations is a quantum mechanical procedure and the rules that govern the interaction of two configurations of different energies are derived from perturbation theory.[1] It turns out that these rules are similar to those governing the interaction of orbitals, described in Chapter 2.

For two configurations, Ψ_A and Ψ_B, whose energies differ significantly, the stabilization energy SE, induced in the low-energy configuration, is described by

$$SE = B^2/\Delta E \tag{3.8}$$

where B, the interaction parameter, is defined by

$$B = \langle \Psi_A | \mathcal{H} | \Psi_B \rangle \tag{3.9}$$

The evaluation of B is often complex, and in many cases is only likely to be accessible by computational means. However, in at least one situation its evaluation is fairly straightforward. If the two configurations that interact just differ in the position of a single electron (and this is very often the case), then the form of B can be greatly simplified, and given an orbital interpretation. In such a case B will equal the resonance integral of the two orbitals that differ in one-electron occupancy.[2]

As an example, consider the interaction of $R \cdot \ \cdot X$ and $R^+ \ :X^-$. For these two configurations B is given by

$$B = \langle R \cdot \; \cdot X \,|\, \mathfrak{IC} \,|\, R^+ : X^- \rangle \tag{3.10}$$

Since these two configurations differ in the position of one electron (an electron has been shifted from ϕ_R, the AO on R, to ϕ_X, the AO on X), then B is approximately equal to the resonance integral of these two orbitals. Thus B may be represented by

$$B = \langle \phi_R \,|\, \mathfrak{IC} \,|\, \phi_X \rangle \tag{3.11}$$

Since the resonance integral between ϕ_R and ϕ_X is related to the corresponding overlap integral, this expression may be further simplified to

$$B = KS_{RX} \tag{3.12}$$

where S_{RX} is the overlap integral between ϕ_R and ϕ_X, and K is an energy constant.

What this means, therefore, is that the stabilization of the ground state, SE, brought about by the mixing of the two configurations, is roughly proportional *to the square of the overlap integral of the two orbitals that differ in one-electron occupancy, and inversely proportional to the energy gap between the two interacting configurations.*

The mixing parameter λ, which measures the extent to which the high-energy configuration mixes into the ground configuration to generate the ground state, is also described by perturbation theory, and is given by

$$\lambda = B/\Delta E \tag{3.13}$$

Thus the extent of mixing is proportional to the overlap and inversely proportional to the energy gap.

In the above example the interacting configurations were of very different energies and the perturbation equations (3.8) and (3.9) were therefore applicable. When the two interacting configurations are degenerate, perturbation theory is no longer valid; interaction of the second configuration with the first configuration can no longer be considered as a perturbation. For the mixing of degenerate configurations, the stabilization energy is given by

$$SE = B \tag{3.14}$$

and the two states obtained through configuration mixing, Ψ_g and Ψ_e, are defined by

$$\Psi_g = \frac{1}{\sqrt{2}}(\Psi_A + \Psi_B) \tag{3.15}$$

$$\Psi_e = \frac{1}{\sqrt{2}}(\Psi_A - \Psi_B) \qquad (3.16)$$

where Ψ_A and Ψ_B are the two interacting configurations. Note, as before, that the equations that govern the interaction of degenerate configurations [Eqs. (3.14–3.16)] are identical in form to those governing the interaction of degenerate orbitals [see Eqs. (2.47) and (2.48)]. They will prove to be of value in our description of curve-crossing diagrams.

Symmetry considerations apply to configuration mixing just as they do to orbital mixing. Just as orthogonal orbitals are precluded from mixing by symmetry, configurations of different symmetry also cannot mix. The symmetry mismatch may be due to either different *spatial* or *spin* characteristics. Thus two configurations that differ in the location of a single electron will not mix if the two orbitals that differ in the electron occupancy are orthogonal. Likewise, two configurations, one singlet and triplet, for example, cannot mix due to their different spin properties.

3.5 THREE-ELECTRON BONDS

While it is true that the covalent bond pair is the most common of all chemical bonds, three-electron bonds between two atoms exist in many species. However, the importance of three-electron bonds extends beyond the specific compounds that contain such bonds. In the course of many chemical reactions, three-electron interactions often play a hidden role in governing reactivity. For example, as we will discuss in Chapter 4, a simple S_N2 reaction between a nucleophile N and an aliphatic substrate $R-X$, generates three-electron interactions in the transition state, *despite* their absence in reactants and products. So understanding the three-electron interaction is important, not just for questions of structure, but for reactivity as a whole. An MO description of the three-electron bond, in terms of a doubly occupied σ orbital and a singly occupied σ^* orbital, was described in Section 2.2.4.

The VB description of the three-electron bond in He_2^+ was first described by Pauling[3] in 1933 and over 60 years later that description remains essentially unchanged. According to Pauling, the three-electron interaction in $A\!\mathbin{\overset{\text{\tiny$\cdot$}}{-}}\!B$ is governed by the interaction of two VB configurations, A: ·B and A· :B. For example, in He_2^+ the two VB configurations are pictorially represented by

$$He: \cdot He^+ \qquad {}^+He \cdot :He$$

3.1   **3.2**

Structures **3.1** and **3.2**. Interestingly, the wave functions represented by **3.1** and **3.2** are actually *repulsive* with respect to He· · ·He approach. The repulsive interaction is due to *exchange repulsion* between the odd electron and the electron of the same spin in the adjacent electron pair. Nevertheless, despite the intrinsic repulsive nature of the three-electron interaction, the three-electron bond can be stabilizing. This is due to the interaction that takes place between the two three-electron forms **3.1** and **3.2,** and is illustrated in the configuration interaction diagram of Fig. 3.4. The negative combination is actually the one that leads to formation of a stable three-electron bonded structure (whose bond strength is estimated to be 50 kcal mol^{-1})^4 and which can be represented by

$$\Psi_g(He_2^+) = (He:\ \cdot He^+) \overset{(-)}{\longleftrightarrow} (^+He\cdot\ :He) \qquad (3.17)$$

while the positive combination corresponds to some high-energy repulsive state. In other words, the resonance stabilization brought about by the mixing of the two resonance forms **3.1** and **3.2** in the ground state, more than compensates for the repulsive nature of the three-electron interaction that is present in each of these forms.

The above description makes it clear that three-electron bonds will show maximum stabilization when the two configurations that contribute to the ground-state wave function are degenerate (when the two VB forms will contribute equally to the state wave function). This explains why radical cations of diamines and disulfides are quite stable and have been experimentally observed.^{4,5} The 4.4.4 bicyclic diamine **3.3** is readily oxidized, even by air, to give the corresponding cation radical **3.4.** Similarly, oxidation of 1,5-dithiacyclooctane leads to the formation of a long-lived cation radical **3.5.** Computational studies reinforce these ideas; the strengths of three-electron bonds in diamines and disulfides have been calculated in $[H_3N \dot{-} NH_3]^+$ and $[H_2S \dot{-} SH_2]^+$ to be 33.0 and 26.5 kcal mol^{-1}, respectively.^{6} Thus we see that the stability of these odd-electron species can be understood in VB terms through

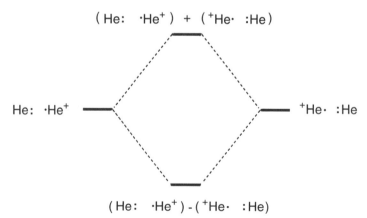

Fig. 3.4. Configuration-mixing diagram showing the mixing of degenerate VB configurations (He: $\cdot He^+$ and $^+He\cdot$:He), to generate ground and excited states of the three-electron bond in He_2^+.

the degenerate mixing of two three-electron VB configurations, such as **3.1** and **3.2.**

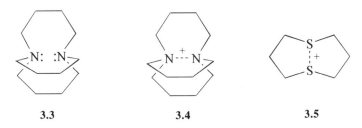

3.3	**3.4**	**3.5**

The oxygen molecule, which has a triplet ground state, is often portrayed as an example of the failure of VB theory, since a simple Lewis structure for O_2 **3.6** suggests that the ground state would be a singlet. However, a triplet ground state *can* be understood with the aid of VB structures, if the oxygen lone-pair orbitals are considered explicitly. Thus O_2 can be viewed as a hybrid of the two VB structures **3.7** and **3.8,** which possess two three-electron bonds, each of which contributes about 30 kcal mol^{-1} to the oxygen–oxygen bond strength.[7] In contrast, the singlet representation for O_2, shown in **3.9**, possesses a regular π bond in the plane of the paper. However, in addition to the stabilizing π bond there is a *destabilizing four-electron* interaction between the two lone-pair orbitals that are now forced into the same plane (perpendicular to the paper). It is this four-electron destabilizing interaction that makes the alternative three-electron bonded structure more stable.

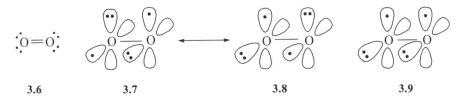

3.6	**3.7**	**3.8**	**3.9**

What happens when the two configurations that contribute to the structure of the three-electron bond are *nondegenerate*? In these cases the three-electron bond may not be stable. Consider the case of the radical anion of an alkyl halide $(R \dot{-} X)^-$. The corresponding interaction diagram is illustrated in Fig. 3.5.

The nature of the three-electron bond will be governed by the interaction of the two contributing configurations

$$R\cdot\ :X^- \qquad R:^- \ \cdot X$$

3.10	**3.11**

For a regular alkyl group, for example, $R = CH_3$, the $(R \dot{-} X)^-$ species is unstable. Because of the large energy gap between the two interacting config-

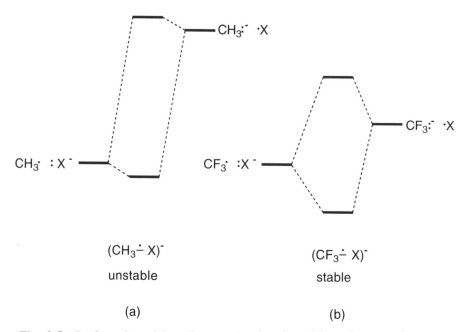

$(CH_3 \dot{-} X)^-$ $(CF_3 \dot{-} X)^-$

unstable   stable

(a)   (b)

Fig. 3.5. Configuration-mixing diagram showing the mixing of VB configurations $R \cdot\ :X^-$ and $R:^-\ \cdot X$, to generate state descriptions of $(R \dot{-} X)^-$. (a) For $R = CH_3$, $R:^-$ is relatively unstable, the state is described primarily by $CH_3 \cdot\ :X^-$ so $(CH_3 \dot{-} X)^-$ dissociates into $CH_3 \cdot$ and $:X^-$. (b) For $R = CF_3$, $R:^-$ is relatively stable, the state is described by the hybrid $CF_3 \cdot\ :X^- \leftrightarrow CF_3:^-\ \cdot X$, so a stable three-electron bond is generated.

urations there is little mixing of **3.11** into the state wave function and therefore there is little stabilization. As a result the repulsive nature inherent in the three-electron configuration **3.10** leads to spontaneous rupture of the three-electron bond. For this reason attempts to generate the radical ion species by the reduction of methyl halides, leads directly to cleavage of the $R-X$ bond,[8] as shown in Eq. (3.18).

$$(R \dot{-} X)^- \rightarrow R \cdot + :X^- \tag{3.18}$$

For $R = CF_3$ the situation is different (Fig. 3.5b). Due to the electron-withdrawing power of the CF_3 group, R^- is stabilized in comparison to a regular alkyl anion so that for CF_3X the energy gap between **3.10** and **3.11** is reduced. This leads to increased mixing of **3.11** into **3.10,** a larger stabilization energy, and hence the formation of stable three-electron bonded species. Indeed, radical anions, such as $[CF_3 \dot{-} I]^-$, $[CF_3 \dot{-} Br]^-$, and $[(CF_3)_3C \dot{-} I]^-$, where the R group is relatively electronegative, have been observed experimentally.[9]

3.6 RELATIONSHIP BETWEEN VALENCE BOND CONFIGURATION MIXING AND RESONANCE THEORY

We already pointed out that resonance theory and the VB configuration-mixing procedure are closely related. Even though both derive directly from VB theory, resonance may be thought of as a simplified version of configuration mixing. Resonance, introduced by Pauling and Wheland[10] some 60 years ago, is used mainly to explain the structure of delocalized species such as the enolate ion and benzene, where one Lewis structure cannot properly describe the electronic distribution and energy of the molecule. Resonance theory considers these molecules in terms of a contribution from several "resonance hybrids".

Configuration mixing is a more formal application of VB theory.[11] It is the quantum mechanically governed procedure whereby an improved state function is generated from the mixing of electronic configurations and is distinguished from resonance mixing in several points. First, configuration mixing leads to both in-phase *and* out-of-phase combinations, that is, to a description of ground *and* excited states, while resonance theory normally considers just the in-phase combination. Second, configuration mixing, being closer to formal quantum mechanics, is governed by rules that enable the procedure to be carried out in a quantitative, or semiquantitative manner. Resonance theory, in contrast, is normally applied as a qualitative method.

Third, and most importantly, the concept of configuration mixing is a general one and is not restricted to well-defined species (i.e., stable molecules and intermediates) in the way that resonance theory is normally applied. Configuration mixing may be utilized to build up an energy profile for a reaction system of several molecules. *This means that not just molecular ground states, but reaction transition states (and excited states) as well,* may be analyzed and characterized according to their contributing VB configurations. Consequently, the method provides a useful tool for understanding reactivity. The way qualitative energy profiles for different reactions may be built up from VB configurations is described in Chapter 4.

REFERENCES

1. For a detailed treatment of the rules that govern configuration mixing see S. S. Shaik. In *New Theoretical Concepts for Understanding Organic Reactions*, Vol. C267, J. Bertran and I. G. Csizmadia, Eds., Kluwer Publications, Dordrecht, The Netherlands, 1989.

2. L. Salem, C. Leforestier, G. Segal, and R. Wetmore. *J. Am. Chem. Soc.* **97,** 479 (1975); N. D. Epiotis. *Theory of Organic Reactions.* Springer-Verlag, Heidelberg, 1978.

3. For a discussion of the three-electron bond see L. Pauling and E. B. Wilson, Jr., *Introduction to Quantum Mechanics*, McGraw-Hill Book Co., New York, 1935.

4. B. Kirste, R. W. Alder, R. B. Sessions, M. Bock, H. Kurreck, and S. F. Nelsen, *J. Am. Chem. Soc.* **107,** 2635 (1985).

5. M. Gobl, M. Bonifacic, and K-D. Asmus, *J. Am. Chem. Soc.* **106,** 5984 (1984); W. K. Musker, T. L. Wolford, and P. B. Roush, *J. Am. Chem. Soc.* **100,** 6416 (1978).

6. P. M. W. Gill and L. Radom. *J. Am. Chem. Soc.* **110,** 4931 (1988).

7. W. A. Goddard, T. H. Dunning, Jr., W. J. Hunt, and P. J. Hay. *Acc. Chem. Res.* **6,** 368 (1973).

8. R. N. Compton, P. W. Reinhardt, and C. D. Cooper. *J. Chem. Phys.* **68,** 4360 (1978); M. C. R. Symons. *Pure Appl. Chem.* **53,** 223 (1981); J. W. Raymonda, L. O. Edwards, and B. R. Russell. *J. Am. Chem. Soc.* **96,** 1708 (1974). For a recent review on the problem of $(R \doteq X)^-$ dissociation, see J-M. Saveant, *Adv. Phys. Org. Chem.* **26,** 1 (1990).

9. A. Hasegawa and F. Williams. *Chem. Phys. Lett.* **46,** 66 (1977); J. T. Wang and F. Williams. *J. Am. Chem. Soc.* **102,** 2860 (1980).

10. L. Pauling and G. W. Wheland, *J. Chem. Phys.* **1,** 362 (1933).

11. For further reading on the VB configuration-mixing approach see A. Pross, *Adv. Phys. Org. Chem.* **21,** 99 (1985); S. S. Shaik. *Prog. Phys. Org. Chem.* **15,** 197 (1985); A. Pross and S. S. Shaik. *Acc. Chem. Res.* **16,** 363 (1983); S. S. Shaik, H. B. Schlegel, and S. Wolfe. *Theoretical Aspects of Physical Organic Chemistry. The S_N2 Mechanism*, Wiley-Interscience, New York, 1992, Chapter 3; S. S. Shaik and P. C. Hiberty. In *Theoretical Models for Chemical Bonding*, Vol. 4, Z. B. Maksic, Ed. Springer-Verlag, Berlin, 1991.

4

BUILDING UP REACTION PROFILES USING VALENCE BOND CONFIGURATIONS

4.1 REACTION PROFILES

In order to represent the reaction pathway for a given process, we need to discuss the concept of a *reaction profile* or *energy profile*. Reaction profiles illustrate the way in which the energy of the reacting system changes as a function of the reaction coordinate. The term energy is used here in a general way and may refer to any one of the different thermodynamic energy measures: enthalpy, H; free energy, G; or internal energy, E. The reaction coordinate is a general term that represents the nuclear reorganization; that is, the *geometric* change in the position of the atoms (relative to one another) that takes place as reactants are converted into products.

Two examples of such profiles are illustrated in Fig. 4.1. Figure 4.1a illustrates an *elementary* reaction, in which reactants R are converted into products P in a single step, while Fig. 4.1b represents a *multistep* (in this case a two-step) reaction that proceeds through some high-energy intermediate I. Both profiles exhibit a characteristic that is very general: The conversion of reactants into products leads the system through at least one high-energy structure that is a point of maximum energy along the reaction coordinate. This special point is termed the *transition state* (labeled $TS^{\ddagger}$). Thus the profile in Fig. 4.1a has one transition state, while that in Fig. 4.1b has two. In recent years there has been a tendency to use the term "transition structure," rather than "transition state," since, strictly speaking, the latter is not a state (for a chemical definition of the term, see Chapter 1). However, in view of the widespread use that the original term enjoys (and the awkwardness of phrases such as "the structure of the transition structure" when it is desired to emphasize the structural aspects

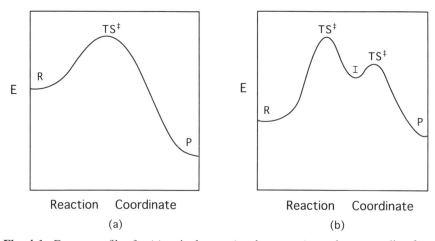

Fig. 4.1. Energy profiles for (a) a single-step (or elementary) reaction proceeding from the reactants R, through the transition state TS‡, and down to products P. (b) A two-step reaction that proceeds through some high-energy intermediate species I.

of the transition state), we will continue to use the term "transition state" in this book.

A central question in organic reactivity is What factors govern the form of the reaction profile? This is a key question since it is the form of the profile that determines the height of the energy barrier, as well as the possible formation and energy of reaction intermediates. In fact, we might ask Why do reactions have a barrier at all? Why is it that in most exothermic reactions, the reactants do *not* spontaneously convert to products? Why is an activation barrier invariably there, and what are the factors that determine its height? What factors determine whether a reaction will be concerted, or step-wise via one or more reactive intermediates? These questions are at the heart of organic reaction mechanisms, and it is therefore important to have a qualitative way of modeling reaction profiles. The means of building up reaction profiles for any organic reaction is the subject of most of this chapter.

4.2 POTENTIAL ENERGY SURFACE DIAGRAMS

Despite their usefulness as a pictorial representation of reaction pathways, reaction profiles have a major limitation: For most reactions the reaction coordinate is not a well-defined quantity. This is the case because the reaction coordinate incorporates a large number of different molecular bending and stretching motions. A full description of the energy of a system, in terms of each and every internal degree of freedom available to that system, requires a multidimensional surface. An alternative method of depicting reaction pathways is through the use of *potential energy surface* diagrams. These have the advantage of giving added geometric information compared to the reaction profiles, but suffer from the disadvantage of representing the energy of the system less clearly. Let us discuss these diagrams in some detail.

Consider a system of N atoms. Since each atom has three degrees of freedom, representing each of the three perpendicular axes, a system of N atoms possesses $3N$ degrees of freedom. Six of these degrees of freedom specify the absolute position and orientation of the system as a whole—three are *position* coordinates (along the x, y, and z axes), and define the position of the center of mass of the system in space, while three are *rotational* coordinates and define the *orientation* of the system in space. The remaining $3N - 6$ degrees of freedom define the possible *internal vibrations* of the system (i.e., the *relative* position of the atoms with respect to one another). *This means that any molecular vibration can be defined in terms of these $3N - 6$ internal degrees of freedom.* (For a linear system there are $3N - 5$ vibrational degrees of freedom, since it possesses just *two* rotational degrees of freedom. The intended third rotational degree of freedom, rotation about the molecular axis, is not a true degree of freedom, since it does not induce any geometric change in the position of the molecule.) Since in a chemical reaction we are interested in the way atoms change their position with respect to one another, chemical reactivity involves tracking $3N - 6$ degrees of freedom (or $3N - 5$ for a linear system) out of the total $3N$.

It is at this point that we begin to run into difficulty. The fact that a system has $3N - 6$ internal degrees of freedom means that a precise plot of energy as a function of the internal atomic motion rapidly becomes intractable. Consider the reaction

$$I^- + CH_3Br \rightarrow ICH_3 + Br^- \qquad (4.1)$$

This relatively simple chemical system is composed of six atoms. Therefore it possesses 12 internal degrees of freedom ($3 \times 6 - 6 = 12$). Consequently, a complete energetic description of the system requires *13* parameters: 12 for the degrees of freedom, and 1 for the energy of the system. And here lies the problem: A simple graphical representation incorporating this degree of complexity does not exist. By using simple graphical techniques the best we can do is to consider just *two* degrees of freedom and their effect on energy. For this we use potential energy surface diagrams. The potential energy surface diagram for an S_N2 reaction, such as that shown in Eq. (4.1), is illustrated in Fig. 4.2a.

In contrast to the energy profiles of Fig. 4.1 in which the energy of the system is plotted against *one* general, ill-defined parameter (the reaction coordinate), the energy in a potential energy surface diagram is plotted as a function of *two* specific geometric parameters. These are the two coordinates in the plane of the paper shown in Fig. 4.2a (we will see shortly which geometric parameters are chosen). The third coordinate (the energy axis) is perpendicular to the plane of the paper so that the energy of the system is indicated by contour lines, just like those that appear on a topographical map. Any line

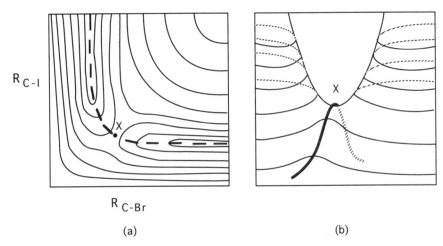

R_{C-I}

X

X

R_{C-Br}

(a)   (b)

Fig. 4.2. (a) Schematic potential energy surface (PES) diagram for the S_N2 reaction, $I^- + CH_3Br \rightarrow ICH_3 + Br^-$. Axes represent C—Br and C—I bond stretching, and the third dimension (energy), which is perpendicular to the plane of the paper, is represented by the contour lines connecting points of equal energy. The reaction coordinate is illustrated by the dashed line through the transition state X. (b) Three-dimensional representation of the transition state region of the energy surface. The transition state X is located on a saddle point.

on the surface joins together points of equal energy, and the spacing between the contour lines signifies the gradient of the surface; closely spaced lines signify a region on the surface in which the gradient is steep, while widely spaced lines signify a region of slight gradient.

In drawing the potential energy surface for the reaction of Eq. (4.1), we may choose, in principle at least, any two of the 12 geometric variables. However, for the diagram to be useful, the two most important geometric parameters need to be chosen. Since the above reaction involves substitution of Br by I, the two parameters that change most drastically during the course of the reaction are the C---I and C---Br bond distances. So these are the two parameters that are chosen for this reaction; other parameters, such as the C—H bond distance and various bond angles, are presumed less important and are not considered. Accordingly, for the S_N2 process of Eq. (4.1), the reaction pathway is indicated by the dashed line in Fig. 4.2a. The reactants lie in the upper left-hand corner of the diagram (consistent with a short C—Br bond and a long C—I bond), and the products lie in the lower right-hand corner. As the reaction proceeds, the C—I bond length decreases while the C—Br bond length increases. At the point marked X, the system reaches its highest energy, so this point is the transition state of the reaction. A three-dimensional representation of the transition state region is illustrated in Fig. 4.2b. A special characteristic of the transition state is that, while it constitutes an energy maximum *along* the reaction coordinate, it represents a local minimum for any geometric change *perpendicular* to the reaction coordinate. In other words, the transition state is located at a *saddle point.* This characteristic of the transition state is a prerequisite, since, if it were *not* a minimum in a direction perpendicular to the reaction coordinate, this would mean that by avoiding this particular point on the surface, an alternative pathway of lower energy would be available to the system. However, such a pathway does not exist (if it did, *it* would become the preferred pathway). So the transition state, by definition, is the highest point that the reaction passes through and defines the lowest energy pathway from reactants to products.

Finally, it should be noted that the profile obtained by taking a cross section of the energy surface (i.e., by cutting through the surface along the reaction coordinate path), would lead to the energy profile of Fig. 4.1a. Thus the two means of representing reaction pathways are closely related. In discussions of reaction pathways, the simpler representation (the reaction profile diagrams of Fig. 4.1) is normally used, though in special cases where added information is required, the potential energy surface diagrams may be referred to.

In order to build up a reaction profile for any reaction, what is needed is a means of determining the energy of the system at each point along the reaction coordinate. For a complex reaction this can only be achieved computationally. Even then the computational time required to plot out reaction pathways, and to measure the energies of transition states, is very significant, exceeding by far the time required to compute the structure and energy of ground-state molecules. However, despite the inherent complexity, it is possible to model reactivity in a simple qualitative and semiquantitative fashion. This is important since it is only through a simpler model that a conceptual framework may be built up that provides intuitive insight into problems of reactivity.

4.3 BUILDING UP A SIMPLE REACTION PROFILE: BOND DISSOCIATION

A useful starting point in trying to understand the factors that are responsible for barrier formation in organic reactions is to explore the simplest of all chemical reactions: the making or breaking of a single bond. Apart from being the most elementary chemical process, the reaction coordinate for such a process is uniquely simple since it involves just one degree of freedom—the distance between the two atoms or groups that form the bond. So in this case plotting energy against the reaction coordinate does not lead to the difficulties associated with more complex reactions, where the reaction coordinate incorporates within it a multitude of geometric changes. But more importantly, we will discover that the principles that govern barrier formation in this simplest of processes, will carry over directly into more complex reactions. So let us consider two prototype reactions in this way: dissociation of the covalent bond in an alkyl halide R—X and dissociation of the ionic bond in LiF, in both the gas phase and in solution.

4.3.1 R—X Bond Dissociation in the Gas Phase

In Section 3.1 we saw that interacting the two configurations that describe the R—X bond (R· ·X and R$^+$:X$^-$) generates better descriptions of both ground and excited states of the bond. In that case the interaction between the two configurations was carried out at a fixed geometry: the R—X bond length. *We can now convert this interaction diagram at fixed geometry, into a reaction profile for R—X dissociation, by introducing a reaction coordinate* (i.e., *by changing the R—X bond distance*).

The way this is done is by plotting the energy of each of the two configurations as a function of the R---X bond distance. In doing so we generate two *configuration curves*. Interaction of these configuration curves generates two *state curves* and it is these two state curves that represent ground and excited energy profiles for R—X dissociation. The procedure is illustrated in Fig. 4.3.

As can be seen in the figure, each of the two configuration curves, indicated by the solid lines, goes through a point of minimum energy when attractive forces are overcome by repulsive ones. The R· ·X curve is attractive at long distances because of the Heitler–London covalent binding (Section 1.4), while the ionic curve, R$^+$:X$^-$, is attractive at long distances due to the electrostatic attraction between the two opposite charges. If, however, the two groups R

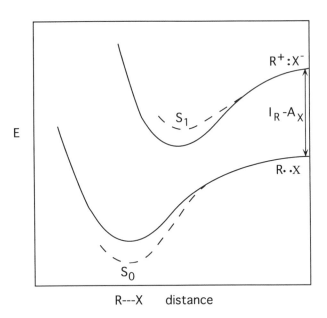

Fig. 4.3. Building up a reaction profile for R—X dissociation in the gas phase by the interaction of the two configuration curves, R· ·X and R$^+$:X$^-$ (firm lines), to generate two state curves, S_0 and S_1 (dashed lines). Dissociation of the ground state S_0 leads to homolytic fission of the bond to give R· and X·, while dissociation of the excited state S_1 leads to heterolytic fission to give R$^+$ and X$^-$.

and X approach one another too closely, then strong electron–electron as well as nuclear–nuclear repulsion set in and the energy of both configurations rises sharply. At large R---X distances the energy gap between the two curves is given by $I_R - A_X$, the energy required to transfer an electron from R to X, to generate R^+ and X^-.

Note that at any point along a configuration curve the electronic structure remains the same; *this is a key feature that characterizes any configuration curve—there is no electronic reorganization as the geometry changes*. This is true by *definition*; the feature that defines an electronic configuration is the way in which the electrons are distributed among the various atoms and is independent of the system geometry. Another way of saying this is that each point on the configuration curve is described by the same electronic wave function.

We saw in Fig. 3.1 that interaction of the two configurations at a fixed geometry led to configuration mixing and the generation of two states. The same procedure can now be carried out here, except that the interaction between the two configurations will now take place *at each point along the R---X dissociation axis*. The two state curves that result (S_0 and S_1) are illustrated by the dashed lines of Fig. 4.3. We see that the ground-state curve, labeled S_0, is *stabilized* in comparison to the R· ·X curve, while the excited-state curve S_1 is *destabilized* relative to the ionic curve, by analogy with what was observed in Fig. 3.1, where the interaction was at a fixed geometry. Note that at infinity, the two state curves merge with the configuration curves, since at infinity there is no mixing between the two configurations. (The degree of mixing of the two configurations is governed by the overlap between the two orbitals ϕ_R and ϕ_X (see Section 3.4). At infinity these orbitals do not overlap, so the two configurations do not mix). Each point on the ground-state curve S_0 is described by the wave function shown in Eq. (3.1), while each point on the excited-state curve S_1 is described by the wave function expressed in Eq. (3.3). Since the two mixing parameters λ and λ' are less than 1 (though they vary in value along the reaction coordinate), we can say that the ground-state curve S_0 is primarily covalent in character, while the excited-state curve is primarily ionic.

This simple picture is in itself a source of interesting information. For example, the diagram in Fig. 4.3 predicts that the stretching of the ground-state R—X bond will lead to *homolytic* fission to yield two radicals R· and X·, while the excited state is predicted to dissociate heterolytically into R^+ and X^- ions. Conversely, combination of the two ions is predicted to generate the *excited* state of the bond (rather than the ground state), and that this excited species would then drop down to the ground-state surface (presumably by some radiative or radiationless transition).

The ground-state behavior is, of course, commonly observed; gas-phase homolytic fission is a routine process. Attempting, however, to generate excited states from gas-phase ion combination is not straightforward; in the gas phase there are problems of energy dissipation (two colliding ions in the gas phase have no means of dissipating the bonding energy, so the internal energy is absorbed by the ions flying apart again). In solution, where energy dissipation can occur, the configuration description is quite different and is discussed separately (Section 4.3.3).

The above example characterizes a system in which the configuration curves do not cross. A very different situation evolves when the configuration curves *do* cross. To discuss this situation we consider the case of LiF dissociation.

4.3.2 LiF Bond Dissociation in the Gas Phase

While the reader might consider that dissociation of the ionic bond in LiF is not quite the archetypal organic reaction, and therefore somewhat out of place in a text on organic reactivity, we will discover that the principles associated with the simple process of stretching an ionic bond have far-reaching consequences for all chemical reactions. In fact, as we will subsequently discuss, there are remarkable similarities between this process and many of the basic polar reactions of organic chemistry, such as nucleophilic addition and substitution, electrophilic addition, etc. We will therefore discuss this elementary reaction in some detail since in the process we will encounter the central concept of the *avoided crossing* for the first time in this text.

An energy plot showing ionic $(Li^+ :F^-)$ and covalent $(Li \cdot \cdot F)$ configuration curves as a function of the Li---F interatomic distance in the gas phase has been obtained from ab initio calculations[1] (see Fig. 4.4a). The configuration curves are indicated by the solid lines.

In contrast to the situation we discovered for R—X dissociation, the two LiF configurations curves (ionic and covalent) *cross* one another. At infinity the covalent configuration is lower in energy than the ionic one because the transfer of an electron from Li to F to give Li^+ and F^- requires the input of about 2 eV (defined by $I_{Li} - A_F$, where $I_{Li} = 5.4$ eV and $A_F = 3.4$ eV). However, as the Li---F distance decreases, the energy of the ionic configuration decreases steeply, due to the electrostatic attraction between the ions, and drops *below* the covalent curve in a deep minimum. Let us now consider the significance of this crossing of the two configuration curves.

In order to specify the energy of the LiF molecule for any given Li---F distance, in both its ground and excited states, we need to generate the two state curves for Li---F approach. This is done by interacting the two configuration curves, as discussed in Section 4.3.1 for the R—X bond. When the energy gap between the two interacting configurations is large (i.e., at Li---F distances $< \sim 5$ Å) and when the Li---F distance is large (i.e., $> \sim 8$ Å), there is essentially no interaction between the configurations (i.e., the mixing parameter λ is small). So, in these two regions the state curve wave functions are well represented by the configuration curve wave functions. However, in the crossing region (where Li---F is 5–8 Å), the energy gap is small, and hence the configurations do interact. Therefore, in this region the state curves will diverge most significantly from the configuration curves. This is shown by the dashed lines of Fig. 4.4a. Thus the *lower* dashed line of Fig. 4.4a represents the *stabilizing* combination of the two configurations (i.e., the ground-state

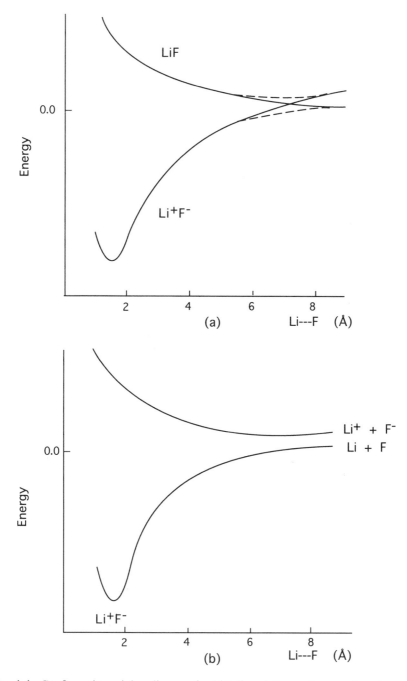

Fig. 4.4. Configuration-mixing diagram for LiF dissociation in the gas phase based on ab initio calculations. (a) Firm lines show the ionic ($Li^+ :F^-$) and covalent ($Li\cdot \ \cdot F$) configuration curves and the dashed lines indicate the region of avoided crossing due to strong configuration mixing. (b) Ground- and excited-state curves for LiF dissociation obtained from the mixing of the configuration curves. [Adapted from L. R. Kahn, P. J. Hay, and I. Shavitt, *J. Chem. Phys.* **61,** 3530 (1974)].

combination), while the *upper* broken line represents the *destabilizing* combination (i.e., the excited-state combination). If we now redraw Fig. 4.4a, with just the two state curves (i.e., leaving out the two configuration curves), we obtain the diagram of Fig. 4.4b and we observe an *avoided crossing*. The significance of the term is clear: whereas the two LiF configuration curves cross one another, as shown in Fig. 4.4a, the state curves do *not* cross; for the state curves the crossing is *avoided* (Fig. 4.4b). The state curves enable us to specify both the energy and electronic structure of LiF for any Li---F distance, in both ground and excited states.

What is the electronic description of LiF at any point along the ground-state curve for LiF dissociation? In contrast to the ground-state curve for $R-X$ dissociation, which is dominated at all geometries by $R\cdot\ \cdot X$, the character of the LiF dissociation curve changes dramatically as a function of the reaction coordinate. At short Li---F distances ($< \sim 5$ Å), the state wave function is described by $Li^+ : F^-$, while at large distances ($> \sim 8$ Å), the state wave function is described by $Li\cdot\ \cdot F$. For the region in between, the wave function is a linear combination of the two configurations

$$\Psi_{LiF} = C_1(Li\cdot\ \cdot F) + C_2(Li^+ : F^-) \qquad (4.2)$$

At the crossing point, where the two configurations are isoenergetic, the coefficients C_1 and C_2 are about equal, indicating that the two configurations contribute about equally to the state wave function. The calculated dipole moment of ground-state LiF supports this expectation. At short Li---F bond distances ($< \sim 6$ Å), the dipole is large, reflecting the ionic character of the bond. However, at long bond distances ($> \sim 7$ Å) the dipole moment drops to almost zero reflecting the atomic character of LiF. As expected, the excited state of LiF exhibits the opposite dipole moment behavior. At short distances calculations reveal the excited state to have a low dipole moment, reflecting the dominant contribution of the covalent configuration to the state wave function, while at large bond distances the dipole moment is calculated to be large, reflecting the dominant contribution of the ionic configuration.

What happens therefore when a Li atom approaches a F atom? In order to answer this, we need to follow the LiF ground-state curve of Fig. 4.4b from infinity to the equilibrium LiF bond distance. Energetically we see immediately that there is no barrier to LiF bond formation—the curve is downhill all the way. Electronically, however, there is a significant change that takes place during LiF bond formation. The electronic character of the system is revealed by inspection of the configuration curve diagram (Fig. 4.4a). As already noted, at large distances the atomic configuration (LiF) is more stable than the ionic one ($Li^+ F^-$), so at infinity, the LiF ground state consists of the two atoms. However, as the interatomic distance decreases, the energy gap between the two configurations decreases, and their energies become identical at the crossing point. *Thus, at the crossing point there is no energetic cost to the transfer of an electron from Li to F.* At even closer Li---F distances it is now the *ionic* configuration that is the more stable one. What this means is that as Li approaches F, an electron transfer takes place at the Li---F interatomic distance, defined by the configuration crossing point. This electron transfer occurs at the crossing point, because it is here that the energy required to transfer the electron from Li to F ($I_{Li} - A_F$) is provided by the electrostatic stabilization energy obtained from the close proximity of a cation to an anion (defined by $-e^2/r$). Indeed, to first approximation, one can predict at what interatomic distance electron transfer will take place from this equality. The relationship is expressed by

$$I_M - A_X \approx e^2/r \qquad (4.3)$$

where M is some metal donor, and X some nonmetal acceptor. The relationship is just an approximate one, since the only energetic interaction between M and X is assumed to be electrostatic. (Interactions between atomic M and X within the covalent configuration M· ·X are ignored, for example.)

The physical significance of the relationship expressed in Eq. (4.3) is simple, yet fundamental: *The electronic arrangement (or in our terminology—configuration) adopted by an atomic system in its ground state, is just the one that is most stable for the particular geometric arrangement of the atoms in that system.* Thus LiF is atomic at large Li---F distances because it is energetically favored to be atomic in this region. At small interatomic distances LiF is ionic, because it is energetically favored to be ionic at these small distances. This common-sense idea will be an important basis of our entire treatment of barrier formation in organic reactions.

From Eq. (4.3) is is clear that the smaller the energy gap between ionic and covalent configurations at infinity, the earlier the configuration curve crossing will take place. This means that the electron transfer will occur at *large* interatomic distances for donors of low ionization potential and acceptors of high electron affinity. For LiF the electron transfer is calculated to occur at 7.2 Å [obtained from Eq. (4.3) by substituting $5.4 - 3.4 = 2$ eV for $I - A$, multiplying by 1.602×10^{-12}, to convert electronvolts to ergs, and substituting -4.8×10^{-10} for e, the electronic charge in electrostatic units (esu); the result gives r in centimeters.] However, for an even better donor acceptor pair, such as CsCl, the energy gap is just 0.28 eV, so that the distance at which electron transfer is calculated to occur is 51 Å! What this means is that for CsCl, the *entire* ground-state curve is effectively dominated by the ionic configuration. Subsequently, we will see that this simple means of assessing the position of electron transfer will be useful for estimating the transition state structure for certain organic reactions.

4.3.3 R—X Bond Dissociation In Solution

In the previous two sections we discussed R—X and LiF bond dissociation (covalent and ionic) in the gas phase. In this section we will build up a reaction profile for the dissociation of the R—X bond in solution, and see how the effect of solvent can be qualitatively incorporated into the curve-crossing model.[2]

The reaction profile for the gas-phase dissociation of the R—X bond was presented in Fig. 4.3. In order to build up a *solution* reaction profile for the same process, we need to consider the effect of solvent on the two configuration curves from which the reaction profile is constructed. Since the major influence of solvent is to stabilize charged species, to first approximation the effect of solvent will be to stabilize the ionic configuration $R^+ :X^-$ relative to the co-valent configuration $R\cdot \ \cdot X$. An energy plot of the two configurations in solution, as a function of the R---X distance, is shown in Fig. 4.5.

The main feature of Fig. 4.5 is that in contrast to Fig. 4.3, the configuration curves, represented by the two unbroken lines, *now actually cross*. This occurs because solvation has lowered the energy of the ionic curve sufficiently to overcome the gas-phase energy gap between $R^+ :X^-$ and $R\cdot \ \cdot X$ at infinity. [For an S_N1 substrate, $I_R - A_X$ is about 3–4 eV (see Tables 7.1–7.3), while ionic solvation energies (listed in Table 8.1) are in the range of 3–4 eV. Thus for two ions the solvation energy is of the order of 6–8 eV]. Mixing of the configuration curves leads to an avoided crossing of state curves (represented by the two dashed lines). This means that the stretching of the R—X bond *in solution* will lead to *heterolytic* bond fission, rather than the *homolytic* fission that takes place in the gas phase. *As the bond is stretched, the ground-state wave function switches its electronic character from predominantly that in* $R\cdot \ \cdot X$ *to that in* $R^+ :X^-$. Therefore, we see that the general question of whether a bond is likely to break heterolytically or homolytically is conveniently handled in configuration terms.

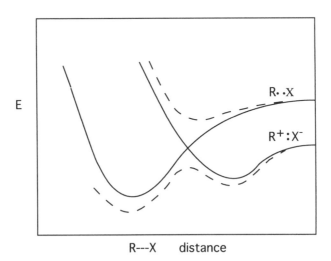

Fig. 4.5. Configuration mixing diagram for R—X dissociation in solution. Firm lines show the crossing of ionic ($R^+ :X^-$) and covalent ($R\cdot \ \cdot X$) configuration curves and the dashed lines indicate the avoided crossing of state curves due to strong configuration mixing.

In the examples discussed above, we considered the different kinds of bond dissociations: gas-phase $R-X$ dissociation exemplifies the situation where a covalent bond breaks up into radicals, solution-phase $R-X$ dissociation represents the case where a covalent bond breaks up into ions, and gas-phase LiF dissociation represents the case where an ionic bond breaks up into radicals. The fourth possibility, in which an ionic bond breaks up into ions, would be represented by the solution dissociation of LiF, and is a simple extension of the above examples.

We may conclude, therefore, that the configuration curve analysis of bond dissociation, described above, allows one to assess the way in which reactivity variations, that is, solvent and substituent changes, are likely to influence the mode in which a bond will break: homolytically or heterolytically.

4.4 THE CURVE-CROSSING MODEL

Most chemical reactions have an activation barrier. While it is true that some processes, such as radical combination, are barrier free, the statement is generally true for chemical reactions where existing bonds are broken and new bonds are formed. Transition state theory teaches us the relationship between the height of the barrier and the rate of the chemical reaction. However, transition state theory does not reveal the cause of the barrier, and the factors that are likely to make it large or small. Indeed it is intriguing that despite the intense effort that has been devoted to the problem of organic reactivity over the past 40 years, little attention has been paid to this most central of questions: Why *do* reactions have a barrier?

Computationally it is an easy matter to demonstrate that simple chemical processes generally involve an activation barrier. The state wave function Ψ is computed for a set of nuclear geometries and then, for each geometry the energy of the system is obtained by solving the Schrödinger equation

$$\mathcal{H}\Psi = E\Psi \tag{4.4}$$

The set of energies plotted as a function of nuclear geometry generates a potential energy surface from which the lowest energy pathway (the reaction coordinate) may be deduced using a variety of computational techniques. In this way a reaction profile, such as the ones illustrated in Fig. 4.1, may be constructed and the barrier height calculated.

If we desire to have some intuitive insight into those factors responsible for generating a reaction barrier, we need to be able to mimic the quantitative procedure described above, by building up a reaction profile qualitatively. The key point here is that the process of constructing the qualitative reaction profile must draw on the quantum mechanical principles on which the quantitative procedure is based. Despite the mathematical complexity associated with this procedure of mapping out a reaction surface, it is possible to mimic the procedure in a qualitatively useful way. The model for describing barrier formation in a chemical reaction is termed the curve-crossing model.[2] We will see that the principles that form the basis of the model derive directly from the bond formation processes described in Section 4.3.

4.4.1 The Nature of the Activation Barrier in Chemical Reactions

Two key configurations that *must* play a dominant role in describing the electronic state for *any* reacting system are the reactant configuration Ψ_R and the product configuration Ψ_P. What characterizes these two configurations? Fortunately, the VB approach enables these two configurations to be specified for any particular reaction by inspection. The procedure is based on the following general statement: *The reactant configuration is just the best VB representation of the reactants while the product configuration is just the best VB representation of the products.* This idea is best illustrated with a typical organic reaction such as the S_N2 process

$$N:^- + R-X \rightarrow N-R + :X^- \qquad (4.5)$$

Ψ_R and Ψ_P for this reaction are represented in pictorial VB form as

$$\Psi_R = N:^- \quad R\cdot \quad \cdot X \qquad (4.6)$$

$$\Psi_P = N\cdot \quad \cdot R \quad :X^- \qquad (4.7)$$

It is apparent that these two VB structures position the key valence electrons in a way that best describes reactants and products, respectively. For example, Ψ_R exhibits the nucleophile lone pair and the $R-X$ bond pair, while Ψ_P exhibits the $N-R$ bond pair and the leaving group lone pair.

The energy of these two configurations is very sensitive to the geometry of the system. For example, Ψ_R is low in energy in the reactant geometry (long N---R distance, short R---X distance) since Ψ_R contains within it the Heitler–London description $R\cdot \quad \cdot X$ of the $R-X$ bond. Of course, in the product geometry (short N---R distance, long R---X distance) Ψ_R is a *high*-energy configuration, since in this geometry the $R\cdot \quad \cdot X$ bonding interaction is absent and has been replaced by a *destabilizing* $N:^- \quad \cdot R$ three-electron interaction. Another way of saying this is that Ψ_R, being a good electronic descriptor of the reactants, is low in energy in the reactant geometry but being a poor electronic descriptor of the products will be *high* in energy in the product geometry. A schematic energy plot of Ψ_R as a function of the reaction coordinate, which shows this increase in energy along the reaction coordinate, is illustrated in Fig. 4.6.

A similar argument applies to the energy of Ψ_P. The function Ψ_P is the main contributor to the state wave function that describes the products, because the electronic arrangement described by Ψ_P is close to the one that actually exists in the product molecules. The energetic behavior of Ψ_P is illustrated in Fig. 4.6, and is the very opposite of Ψ_R. Since Ψ_P contains the Heitler–London

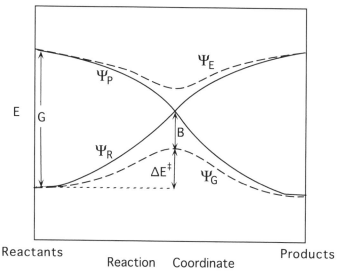

Fig. 4.6. Generalized configuration-mixing diagram for a reaction [Reactants (R) → Products (P)], which shows the crossing of the reactant configuration curve Ψ_R, and the product configuration curve Ψ_P (indicated by the firm lines). The avoided crossing of the two state curves, Ψ_G (ground) and Ψ_E (excited) (indicated by the dashed lines), is obtained from the mixing of the two configuration curves.

description of the new N—R bond that is formed, N· ·R, it is low in energy in the product geometry (short N---R distance) but high in energy in the reactant geometry, where there is no N· ·R bonding and the short R—X distance causes R· :X⁻ three electron repulsion.

The X-type pattern, illustrated in Fig. 4.6 and brought about by the crossing of the two configuration curves, is extremely general and will occur for any chemical reaction in which old bonds are broken and new ones are formed. It must come about because Ψ_R is, by definition, a good descriptor of reactants and a poor descriptor of products, while Ψ_P is the reverse.

Having built up reactant and product configuration curves, let us now generate an energy plot of the ground-state wave function, Ψ_G (G = ground). By doing so we will have built up in a qualitative fashion the reaction profile for the given reaction (yet based on quantum mechanical principles). We discussed in detail (Section 1.7 and Chapter 3) how the mixing of configuration wave functions generates state wave functions. This same mixing procedure may be applied to Fig. 4.6 where the two configuration wave functions Ψ_R and Ψ_P are allowed to interact at each point along the reaction coordinate. As a result, two *state* curves are generated—a lower curve that describes the energy of Ψ_G (the ground-state wave function) and an upper curve that describes the energy of Ψ_E (E = excited) (the excited-state wave function). These are illustrated by the dashed lines in Fig. 4.6. Following Eq. (1.20), which describes how any state wave function is described as a linear combination of configuration wave functions, Ψ_G may be defined as a linear combination of Ψ_R and Ψ_P. Thus

$$\Psi_G = C_1 \Psi_R + C_2 \Psi_P \tag{4.8}$$

At the reaction starting point, where the energy gap between Ψ_R and Ψ_P is large, there is little mixing of the high-energy Ψ_P into Ψ_R and Ψ_G is described largely by Ψ_R ($C_1 \gg C_2$). In this region there is little stabilization so that Ψ_G has essentially the same energy as Ψ_R. At the reaction end the reverse is true; it is now Ψ_P that dominates Ψ_G ($C_2 \gg C_1$), and once again there is little stabilization. Toward the center of the reaction coordinate, however, the situation is quite different. Here Ψ_R and Ψ_P are of similar energies, so that $C_1 \approx C_2$, and the extent of mixing is large. Thus, according to a simple two configuration model, it is in the transition state region that much of the electronic reorganization takes place. Since at the crossing point the two configurations are of equal energies, the electronic description of the transition state may be described by a resonance hybrid of reactant and product configurations, in which the two configurations contribute equally

$$\Psi_{TS} = \frac{1}{\sqrt{2}} \{\Psi_R \leftrightarrow \Psi_P\} \tag{4.9}$$

In the crossing region Ψ_{TS} is *stabilized* in comparison to the crossing point of Ψ_R and Ψ_P by an amount B, where B is defined by Eq. (3.9).

Thus the net result of the interaction of the two configuration curves is to generate an *avoided crossing* of two state curves: a ground-state curve Ψ_G and an excited-state curve Ψ_E. The ground-state curve governs the thermal barrier to reaction, while the excited-state curve may be accessed by the appropriate photochemical excitation. Indeed, with these two curves, one can illustrate the relationship between the ground-state (thermal) reaction pathway leading from reactants to products (Route T) and the excited-state (photochemical) pathway (Route P), though the precise factors that govern the mechanism of decay from the excited-state surface to the ground-state surface require further study.[3] The two routes are illustrated in Fig. 4.7.

The ability to model the ground-state activation barrier for a chemical reaction in the simple way just described, is useful for several reasons. The curve-crossing model (a) provides a means of estimating barrier heights in chemical reactions in terms of established physical parameters, (b) provides a structural and electronic description of the transition state, and (c) provides a means of assessing the effect of substituent and solvent changes on reactivity. Specific examples of these types of applications are discussed in Chapters 9 and 10.

Let us then summarize the essence of the curve-crossing model. A chemical reaction involves both geometric and electronic reorganization. The barrier to a chemical reaction comes about from the way these two reorganizations are

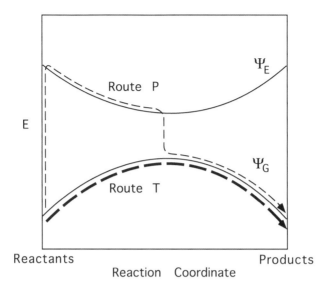

Fig. 4.7. Diagram illustrating thermal and photochemical pathways for a reaction. The thermal pathway follows the ground-state surface (Route T, shown by bold dashed line), while the photochemical pathway involves initial excitation to the excited-state surface, and subsequent decay back to the ground-state surface (Route P, shown by regular dashed line).

coupled to one another. The barrier arises from the need to distort the structure of the reactants in a way that simultaneously *lowers* the energy of the product configuration and *raises* the energy of the reactant configuration to the point that they become *isoenergetic.* Thus it is the geometric changes that bring these two electronic configurations into resonance with one another and facilitates the electronic reorganization (from Ψ_R to Ψ_P). In other words, *geometric distortion is the mechanism that enables the electronic rearrangement to take place.*

4.4.2 Factors Governing Barrier Heights

From the schematic diagram in Fig. 4.6 it is apparent that the barrier height $\Delta E^{\ddagger}$ may be expressed in terms of the initial energy gap G between Ψ_R and Ψ_P, f, a parameter in the range 0–1, which measures the fraction of the initial energy gap that contributes to the barrier height (the magnitude of f is governed by the slopes of the configuration curves), and the quantum mechanical interaction parameter B as follows:

$$\Delta E^{\ddagger} = fG - B \tag{4.10}$$

From Eq. (4.10) it is apparent that a large G value will lead to a large barrier $\Delta E^{\ddagger}$, while a small G value will lead to a small barrier (for given f and B

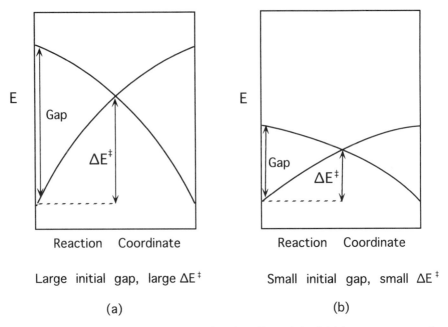

Fig. 4.8. Curve-crossing diagram showing the effect of the initial energy gap on the height of the reaction barrier $\Delta E^{\ddagger}$.

values). This result is illustrated in Fig. 4.8. The effect of f on the barrier height can be seen in Fig. 4.9. Thus the configuration curve slopes shown in Fig. 4.9a lead to a large f and a large barrier, while the configuration curve slopes shown in Fig. 4.9b lead to a small f and a small barrier. Note that changes in reaction exothermicity would affect the barrier through changes in the value of f; more exothermic reactions would have a smaller f value.

In principle, on the basis of Eq. (4.10), absolute barrier heights may be calculated once the magnitudes of f, G, and B, are known. In practice, however, the procedure is not straightforward. While the estimation of G is relatively simple, evaluating absolute values of f and B is much more difficult. This means that the primary utility of Eq. (4.10) is as a *relative* measure of reactivity. Thus in a typical application, the *relative* barriers for two related reactions may be compared, by assessing how the reaction perturbation (e.g., the effect of a given substitutent change) will affect the parameters G and f (B is normally assumed to remain constant), and through them the barrier height. Examples of this type of application, based on changes in G, are described in Chapters 9 and 10.

4.4.3 Nature of the Product Configuration

From the foregoing discussion it becomes clear that the relationship between the reactant configuration Ψ_R and the product configuration Ψ_P is crucial in

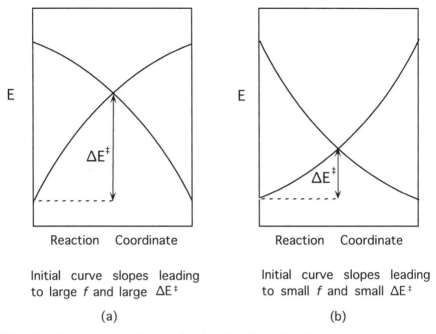

E

$\Delta E^{\ddagger}$

Reaction   Coordinate

Initial curve slopes leading
to large f and large $\Delta E^{\ddagger}$

(a)

E

$\Delta E^{\ddagger}$

Reaction   Coordinate

Initial curve slopes leading
to small f and small $\Delta E^{\ddagger}$

(b)

Fig. 4.9. Curve-crossing diagram showing the effect of configuration curve slopes on the height of the reaction barrier $\Delta E^{\ddagger}$.

governing barrier heights. The relationship between Ψ_R and Ψ_P is particularly important in determining the size of the energy gap G. Therefore some consideration of this relationship, and how it influences the gap size, is now appropriate.

At the reactant geometry, Ψ_P represents an *excited* configuration with respect to Ψ_R. Given that electronic states can generally be approximated by a single configuration, it follows that Ψ_P is likely to be a reasonable representation of some excited state. So a key question to consider is *At the reactant geometry, what is the nature of the excited state that is represented by Ψ_P?* We will see that despite the enormous variety of organic reaction types, organic reactions may be assigned to two basic classes. In the first class, the excitation of Ψ_R to Ψ_P takes place by an *electron shift*, while in the second class, excitation takes place by a *spin shift*. Certain reactions involve both an electron shift and a spin shift.

The electron shift class of reactions is defined as that class of reactions, whose barrier is generated by the avoided crossing of DA and D^+A^- configuration curves; that is, the electronic reorganization that takes place in these reactions is the shift of an electron from a donor molecule D to an acceptor molecule A. DA labels the electronic arrangement before the electron shift, while D^+A^- labels the arrangement after it. The S_N2 reaction [Eq. (4.5)], belongs to this class, because inspection of reactant and product configurations for this reaction [Eqs. (4.6) and (4.7)], reveals that the electronic reorganization required to convert Ψ_R (i.e., N:⁻ R· ·X), to Ψ_P (i.e., N· ·R :X⁻), is just the shift of a single electron from N to X. In fact, the core of basic undergraduate organic reactions, such as nucleophilic substitution and addition, electrophilic substitution and addition, and proton transfer, belongs to this general class of *single-electron shift* processes.

Since this reaction class is characterized by an electron shift, the initial energy gap G is given by

$$G = I_D - A_A, \tag{4.11}$$

where I_D is the vertical ionization potential of the donor, and A_A is the vertical electron affinity of the acceptor. (Vertical, rather than adiabatic energy terms are used since the energy gap is specified at a *frozen geometry—the reactant geometry*. Adiabatic electron-transfer energies are smaller than vertical electron-transfer energies since both donor and acceptor can reorganize structurally to accommodate the electronic change.) *In other words, a key measure of reactivity in this class of reactions may be obtained, simply by consulting tables of ionization potentials and electron affinities.*

Let us now discuss the *spin shift* class of reactions that includes certain radical processes and bond-exchange reactions. In this class of reaction the electronic distribution itself remains unchanged and it is just the electron *spins* that are reorganized. Consider the identity exchange reaction

$$D· + H-H \rightarrow D-H + H· \tag{4.12}$$

The reactant and product configurations for this reaction may be represented by the following structures, in which the electron spins are explicitly labeled:

Reactant configuration	Product configuration
4.1	**4.2**

In the reactant configuration (**4.1**) the electron spins are arranged so that the two electrons in the H−H bond are spin paired, while the odd electron on the D atom and the central H atom are not paired. In the pictorial representation of **4.1** they appear to form a triplet interaction, since both electrons are of the same spin. (This is not strictly true since the covalent bond in H_2 is actually represented by the resonance interaction $\alpha\beta \leftrightarrow \beta\alpha$. The main point here is that these two electrons are *not* coupled into a covalent bond pair. However, the triplet state *is* the closest spectroscopic state to represent the D· · ·H interaction in the reactant configuration, and so is the one that is chosen.)

In the product configuration (**4.2**) the arrangement is reversed; it is now the electrons on the D and central H atoms that are spin paired, while the two electrons on the two H atoms are now uncoupled. Note that in order to convert reactant configuration to product configuration, no electrons need to be shifted, since in both configurations each atom possesses a single electron. What *has* taken place during the conversion is a *spin shift*—the spin arrangement has been modified so that the atoms that enjoy bond coupling are now the D atom and the central H atom, rather than the two H atoms. *The unpaired spin has shifted from the D atom to the right-hand H atom.*

The question now arises What is the nature of the excitation that excites the reactant molecules from the ground state (represented by **4.1**) to the specific excited state, which has the electronic distribution of the products (represented by **4.2**)? The answer: *It is a H—H singlet-to-triplet excitation that best describes the conversion of the reactant configuration to the product configuration.* This occurs because it is the singlet–triplet excitation that leads to uncoupling of the H—H bond pair, while simultaneously facilitating the coupling of the new D—H bond pair. In other words for the atom exchange reaction of Eq. (4.12), the two key configurations in the avoided crossing diagram will be DA and D^3A^*. The $^3A^*$ symbol in D^3A^* signifies that the acceptor molecule (i.e., H—H) has been excited to the triplet state. Thus, for such a reaction, *the initial energy gap G [Eq. (4.10)] is approximated by the singlet–triplet energy gap for the H—H bond—the bond that is to be broken.*[a] This means that reactions in which the bonds to be broken have *large* singlet–triplet gaps will have larger barriers than reactions whose corresponding bonds have *small* singlet–triplet gaps. Such a pattern is in fact observed in radical and pericyclic reactions and is discussed in Chapter 10.

Finally, it should be pointed out that not all reactions in the spin shift class are described by a DA—D^3A^* avoided crossing. In general for this class of reaction, a singlet–triplet excitation is needed for *each* bond that needs to be broken. So, for example, the bond exchange reaction between hydrogen and deuterium molecules in which *two* bonds are broken and two new bonds are

$$
\begin{array}{ccc}
\text{H—H} & & \text{H}\quad\text{H} \\
& \xrightarrow{\hspace{2cm}} & |\quad\ | \\
\text{D—D} & & \text{D}\quad\text{D}
\end{array}
\qquad (4.13)
$$

formed, would be described by the avoided crossing of a DA configuration with a $^3D^*{}^3A^*$ configuration. In this case uncoupling of *both* H—H and D—D bonds is required to generate the two new H—D bonds. Reactant and product configurations for this process are shown in configurations **4.3** and **4.4**.

---

[a] A detailed analysis actually shows that the gap is equal to three quarters of the singlet–triplet excitation because the triplet state is not the precise electronic description of the products, just the spectroscopic state that best describes the products. See Ref. 4 for a detailed description.

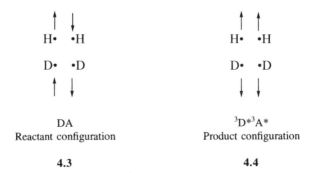

DA

Reactant configuration

4.3

$^3D^{*3}A^*$

Product configuration

4.4

In the reactant configuration, the electron spins are coupled in H—H and D—D bond pairs, while in the product configuration the electrons are coupled in H—D bond pairs. The point is that in the reactant geometry (i.e., short H---H and D---D distances, but long H---D distances), the product configuration represents H—H and D—D excitation to local triplets. In other words in order to facilitate the bond exchange process, *both* of the bonds that need to be broken have to be excited to local triplets so as to generate the electronic configuration of the products. In Chapter 10 we will see that the barriers to cycloaddition reactions are generated by a $DA-^3D^{*3}A^*$ avoided crossing.

Till now we have described concerted reactions whose barrier may be described by the avoided crossing of reactant and product configurations. Let us now consider reactions that require more than two configurations for a proper description, and discuss how we can understand the occurrence of *step-wise* reactions that proceed through some high-energy intermediate.

4.4.4 Effect of Intermediate Configurations

Building up a reaction profile from just two configurations is a useful approximation, and leads to important insights into reactivity problems. However, there are some reactions for which this approximation is inadequate. For such reactions, at least one additional configuration, Ψ_I (I = intermediate), which typically describes the electronic distribution of a potential intermediate, is necessary in order to build up the reaction profile. The energy of the electronic configurations for a generalized reaction, which requires this intermediate configuration for a more accurate description, is plotted in Fig. 4.10. In addition to the X pattern generated by Ψ_R and Ψ_P, a curve, representing Ψ_I, is now included in the diagram. The resultant ground-state curve Ψ_G obtained by the mixing of Ψ_R, Ψ_P, and Ψ_I, is shown as a bold curve.

From Fig. 4.10 it can be seen that Ψ_I is likely to be relatively high in energy at reaction ends compared to Ψ_R and Ψ_P. Why is this so? The configurations Ψ_R and Ψ_P are ground configurations at one end of the reaction coordinate and excited configurations at the other end. However, Ψ_I *is by definition an excited configuration at both reactant and product geometries*. It must be an excited configuration in the reactant geometry, because Ψ_R is the ground configuration

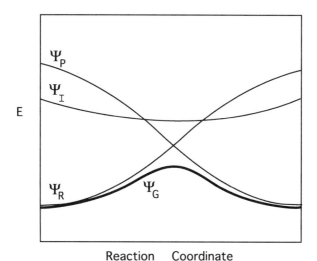

Fig. 4.10. Curve-crossing diagram showing the mixing of configuration curves for reactants (Ψ_R), products (Ψ_P), and intermediate Ψ_I, to generate a single-step ground-state reaction profile, (Ψ_G) (shown by the bold line).

for the reactants so *any* other configuration at this geometry must represent some excitation of the reactants. The same argument applies to the product geometry. In the product geometry it is now Ψ_P that is the ground configuration, so that, here also, any other configuration must represent some excitation of Ψ_P.

The energetic behavior of Ψ_I in Fig. 4.10 reveals that while Ψ_I may contribute only slightly to a description of reactants and products (because of the high energy of Ψ_I at reaction ends, compared to Ψ_R and Ψ_P), it may contribute significantly to a description of Ψ_G in the transition state region. This follows since in the transition state region Ψ_R, Ψ_P, and Ψ_I may well have similar energies. Thus the wave function at the transition state Ψ_{TS}, would be described by

$$\Psi_{TS} = C_1\Psi_R + C_2\Psi_P + C_3\Psi_I \qquad (4.14)$$

where the magnitudes of the coefficients C_1, C_2, and C_3 may result in the substantial mixing of all three configurations Ψ_R, Ψ_P, and Ψ_I, into the state wave function. If one assumes that the extent of mixing of Ψ_I into the transition state wave function does not disturb the equality of the contributions of reactant and product configurations (i.e., $C_1 \approx C_2$), then we may write:[5]

$$\Psi_{TS} \cong N\left\{\frac{1}{\sqrt{2}}[\Psi_R + \Psi_P] + \lambda\Psi_I\right\} \qquad (4.15)$$

where λ is the mixing coefficient of Ψ_I into the transition state wave function and N is the normalization constant that can be shown to equal $(1 + \lambda^2)^{-1/2}$.

Let us return to our example of the S_N2 reaction [Eq. (4.5)] to clarify the concept of an intermediate configuration Ψ_I. For this reaction a possible intermediate configuration Ψ_I (termed the carbocation configuration, for obvious reasons) is represented by

$$\Psi_I = N:^- \ R^+ \ :X^- \qquad (4.16)$$

Accordingly, the transition state for a reaction, in which this intermediate configuration contributes significantly, would be described in the language of resonance as

$$TS = (N:^- \ R\cdot \ \cdot X) \leftrightarrow (N\cdot \ \cdot R \ :X^-) \leftrightarrow N:^- \ R^+ \ :X^- \qquad (4.17)$$

Thus the transition state would reflect the electronic character of reactant, product, *and* intermediate configurations, as indicated in

$$N^{\delta-}\text{---}R^{\delta+}\text{---}X^{\delta-}$$

4.5

that is, partial negative charges on N and X, partial positive charge on R, and partial N---R and R---X bonding.

The formation of an *actual* intermediate along a reaction pathway will occur in those cases where there is substantial mixing of Ψ_I into the ground-state wave function Ψ_G. This can be understood by consideration of Fig. 4.11. In

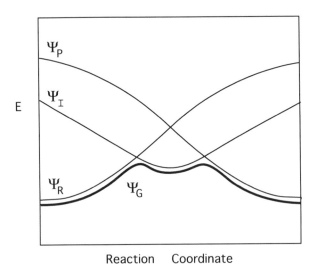

Fig. **4.11.** Curve-crossing diagram showing the mixing of configuration curves for reactants (Ψ_R), products (Ψ_P), and intermediate (Ψ_I) to generate a two-step ground-state reaction profile (Ψ_G) (shown by the bold line).

this case a low lying Ψ_I leads to a *step-wise* reaction pathway. The first step is generated primarily by the avoided crossing of Ψ_R and Ψ_I (with just slight mixing of Ψ_P), while the second step comes about through the avoided crossing of Ψ_I and Ψ_P (with slight mixing of Ψ_R). The local minimum along the Ψ_G curve signifies the formation of an intermediate. The electronic character of this intermediate is governed primarily by Ψ_I, since at that point along the reaction coordinate, Ψ_I is the low-energy configuration, and therefore the one that dominates the ground-state wave function Ψ_G. Thus, in our example of the S_N2 reaction, the intermediate that would be formed would be a nucleophilically solvated ion-pair, whose electronic structure would be represented by the resonance structure of Eq. (4.16).

We see therefore that introducing a third configuration adds flexibility to the configuration-mixing procedure. It provides a means of incorporating electronic character into the ground-state wave function that may be absent in both reactants and products. We will subsequently see that this additional configuration provides a way of understanding a variety of chemical phenomena. Some examples that will be discussed in this fashion include mechanistic variations within a reaction family, anomalous rate-equilibrium relationships, and breakdowns in the reactivity–selectivity principle.

REFERENCES

1. L. R. Kahn, P. J. Hay, and I. Shavitt, *J. Chem. Phys.* **61,** 3530 (1974).

2. For reviews of the curve-crossing model, see A. Pross, *Adv. Phys. Org. Chem.* **21,** 99 (1985); S. S. Shaik. *Prog. Phys. Org. Chem.* **15,** 197 (1985); A. Pross and S. S. Shaik. *Acc. Chem. Res.* **16,** 363 (1983); S. S. Shaik and P. C. Hiberty. In *Theoretical Models for Chemical Bonding*, Vol. 4, Z. B. Maksic, Ed. Springer-Verlag, Berlin, 1991. For a recent monograph on the curve-crossing model that treats the S_N2 reaction in detail, see S. S. Shaik, H. B. Schlegel, and S. Wolfe, *Theoretical Aspects of Physical Organic Chemistry. The S_N2 Mechanism*, Wiley-Interscience, New York, 1992.

3. For recent thoughts on the mechanism of decay of excited states to the ground state, see S. Shaik and C. A. Reddy, *J. Chem. Soc. Faraday Trans.* **90,** 1631 (1994).

4. S. S. Shaik, P. C. Hiberty, J-M. Lefour, and G. Ohanessian, *J. Am. Chem. Soc.* **109,** 363 (1987).

5. S. Shaik, A. Ioffe, A. C. Reddy, and A. Pross, *J. Am. Chem. Soc.* **116,** 262 (1994). See also Ref. 3.

PART B

PRINCIPLES OF PHYSICAL ORGANIC CHEMISTRY

5

PRINCIPLES OF REACTIVITY

5.1 INTRODUCTION

The evolution of physical organic chemistry as a separate area of study within organic chemistry, with its emphasis on problems of structure and reactivity, was a natural consequence of the enormous developments in organic chemistry during this century. Even though the fundamental tenets of chemical behavior were laid down in the 1920s and 1930s with the discovery of quantum mechanics, the relative complexity of organic systems meant that the influence of theoretical chemistry on the development of physical organic chemistry was, till quite recently, rather limited. Certainly, basic theoretical concepts such as the chemical bond, resonance theory, and hydridization, to mention a few, have been central to the development of chemical thinking. However, much of the large body of empirical data that has accumulated has been organized and categorized, rather than explained in terms of basic theoretical principles. The ability to predict a priori in a qualitative way, the rate at which two molecules A and B will react, remains frustratingly out of reach. At best, chemists today are content to grapple with the question Given the rate of reaction between A and B, what will the rate be for the reaction between A and B', where B' is related structurally to B.

The absence of basic theoretical principles to underpin parts of the framework of physical organic chemistry has resulted over the years in the evolution of a large number of empirical rules. Some of these rules have been extremely useful, but others may be less reliable and may need reevaluation. In this chapter we will review the conceptual tools that are commonly applied in physical organic chemistry to probe questions of reactivity.

5.2 CONCEPT OF A REACTION MECHANISM

Consider the reaction

$$A + B \rightarrow Products \tag{5.1}$$

If we want to make some prediction as to how fast that reaction is likely to proceed, or, alternatively, if we *know* the rate of that reaction, but wish to predict the rate of a related reaction

$$A' + B \rightarrow Products \tag{5.2}$$

(e.g., in which a substituent has been introduced into one of the reactants), we need to "know" the *reaction mechanism* for that particular reaction. Knowing the reaction mechanism for a reaction is essential for making any predictions regarding chemical reactivity, and it is for this reason that the study of reaction mechanism has long been at the heart of much chemical research.

What is actually meant by the term "knowing the reaction mechanism"? Bunnett's recent remarks on this question are worth quoting.[1]

> A reaction mechanism is a detailed description of a reacting system as it progresses from reactants to products. It includes identification of all intermediates that are involved, assessment of the characteristics of the transition state(s) through which the reaction progresses, and recognition of the factors that affect reactivity . . .
>
> . . . one cannot expect to know or to prove a reaction mechanism in an absolute sense. The chemist can often reject conceivable mechanisms on the basis of experimental evidence and thereby narrow the field of possibilities, perhaps till only one remains . . . but how can one know whether there remains an inconceivable mechanism that is in full accord with the facts?

We learn from the above statements that the study of reaction mechanism is invariably associated with some uncertainty. The proposition that in science one cannot ever prove a theory—only falsify one—is particularly pertinent to the study of reaction mechanism. A reaction mechanism can be *disproved* much more confidently than it can be *proved*.

At its most basic level, the reaction mechanism involves specifying the number of *elementary steps* from which the overall process is composed, identifying the one that is *rate determining*, and characterizing the *intermediates* that are formed after each of the elementary steps. For example, the statement that the nucleophilic substitution reaction of $R-X$ by Y^- to give $R-Y$ and X^- proceeds through a two-step reaction pathway

$$R-X \xrightarrow{\text{RDS}} R^+ + X^-$$
$$R^+ + Y^- \longrightarrow R-Y \tag{5.3}$$

(where RDS is the rate-determining step) in which the intermediate formation of a carbocation R^+ is rate determining, constitutes a basic description of the reaction mechanism. With just this limited knowledge one could predict fairly confidently that $R'—X$ would react *faster* than $R—X$, in those cases where R'^+ is *more stable* than R^+ (the relative stabilities can normally be predicted, given that the effect of structure on carbocation stability is well understood). Thus, we see that by knowing the reaction mechanism, one may be able to predict the relative rates of the two substitution processes $R—X$ and $R'—X$ with Y^-.

A more complete description of the reaction mechanism also requires *characterization of the transition state* for each elementary step. In particular, characterization of the transition state for the *rate-determining step* is especially important. Consider the simple example of a one-step substitution reaction of $R—X$ by Y^-. Knowing the number of elementary steps (in this case, just one), and knowing the structure of reactants and products would not, in itself, be enough to adequately characterize the reaction mechanism. In this particular case (where no intermediates are formed during the reaction), determining the *geometric* and *electronic structure* of the transition state would be important for a proper description of the reaction mechanism. For example, knowing whether the substitution of X by Y takes place via back-side attack or front-side attack and knowing whether there was a buildup of charge (either positive or negative) on the carbon atom undergoing substitution, would be of considerable importance, if one wished to make a prediction regarding the relative reactivity of two substrates $R—X$ and $R'—X$ toward some species Y^-. Only with this additional information, would one attempt to predict the effect of replacing R by R' on the rate of the reaction. Normally, however, such a detailed description of a complex reaction mechanism is impractical and quite out of reach. A basic description, in which the number and nature of all the reactive intermediates is established, is often quite adequate.

One of the most important tools in the study of reaction mechanism is the tool of *chemical kinetics*. Kinetic studies provide direct information regarding the *energy of the transition state*. Let us describe how this is done.

5.3 THE ARRHENIUS EQUATION

The rate of a chemical reaction is the quantitative measure of chemical reactivity, and its evaluation lies within the domain of chemical kinetics. Many excellent texts on reaction kinetics are available so a detailed treatment of kinetics is not presented here.[2] What we will focus on here is how simple rate data can be interpreted in terms of reaction profiles.

The most direct means of experimentally establishing the height of the energy barrier for a given reaction is to determine the *rate constant k* for that reaction. From simple kinetics we know that the *rate* of a chemical reaction is dependent on the rate constant k and the reactant concentrations. For example, the equation describing the rate of a second-order reaction is

$$\text{Rate} = k\,[A][B] \qquad (5.4)$$

Barrier heights are related to the *rate constants* rather than to the rates themselves because the energy profile for a reaction should be a characteristic of the reaction and *independent* of reactant concentration. If barrier heights were to depend on rates (rather than the rate constants), then a change in reactant concentration would necessarily result in a corresponding change in the energy profile; a unique energy profile would then no longer characterize a particular reaction.

The simplest relationship between reaction rate constant and barrier height is given by the *Arrhenius equation*

$$k = A \exp\left(-E_a/RT\right) \qquad (5.5)$$

where A is termed the *preexponential* or *frequency factor*, E_a is termed the *Arrhenius activation energy* (or activation energy), R is the universal gas constant (equal to 1.987 cal mol^{-1} deg^{-1} or 8.314 J mol^{-1} deg^{-1}), and T is the temperature in degrees kelvin (K). [Since the units of rate are moles per liter per time (mol L^{-1} time^{-1}), it is apparent from Eq. (5.4) that the units of the rate constant k depend on the *order* of the chemical reaction. Thus, it is easy to show that for a first-order reaction the units are time^{-1}, while for a second-order reaction the units are liter per mole per time (L mol^{-1} time^{-1}).]

The basis for the Arrhenius equation is empirical. It derives from the observation that for most reactions there is *a linear relationship between the reciprocal of the temperature and the logarithm of the rate constant*, as observed from a plot of ln k versus $1/T$. This relationship can be expressed as

$$d\,(\ln k)/d(1/T) = \text{constant} \qquad (5.6)$$

Since this has a similar form to the differential equation that describes the effect of temperature on equilibrium constants

$$d\,(\ln K)/d(1/T) = -\Delta H/R \qquad (5.7)$$

we may, by analogy, write Eq. (5.7) as

$$d\,(\ln k)/d(1/T) = -E_a/R \qquad (5.8)$$

where we have simply replaced the equilibrium constant K, by the rate constant k, and the thermodynamic energy term ΔH, by the activation energy term E_a. Integration of Eq. (5.8) then leads to the Arrhenius equation [Eq. (5.5)]. Thus

E_a may be thought of as the minimum energy required by the reactants in order to overcome the activation barrier.

The general form of the Arrhenius equation may be readily justified in another way. The rate of a bimolecular reaction is likely to depend on the number of collisions per second, Z, that take place between the reacting molecules, and some fraction f, which defines the number of successful collisions. Thus,

$$\text{Rate} = fZ \tag{5.9}$$

If we assume that only collisions with an energy greater than some critical energy $E^{\ddagger}$ will be successful, then the fraction of collisions f, whose energy will be greater than $E^{\ddagger}$, is given by the Maxwell–Boltzmann theory of energy distribution as

$$f = \exp\left(-E^{\ddagger}/RT\right) \tag{5.10}$$

Substituting Eq. (5.10) into Eq. (5.9) gives

$$\text{Rate} = Z \exp\left(-E^{\ddagger}/RT\right) \tag{5.11}$$

The analogy between the Arrhenius equation (Eq. 5.5) and Eq. (5.11) is evident.

The experimental determination of the two Arrhenius parameters A and E_a is straightforward and may be obtained by measuring the rate constant k for a reaction at a number of different temperatures. The two Arrhenius parameters E_a and A may be extracted from an experimental plot of $\ln k$ versus $1/T$; the slope of this plot is $-E_a/R$ and its intercept is $\ln A$. From the Arrhenius equation it is a simple matter to demonstrate the well-known rule of thumb that increasing the temperature of a reaction by 10°C roughly doubles the reaction rate. (Of course, the actual increase in rate will depend on the value of E_a—the higher the activation barrier, the greater the dependence of reaction rate on temperature change.)

If the reaction is an elementary process (e.g., as shown in Fig. 4.1a), then the measured E_a is an approximate measure of the barrier height for that process. However, if the reaction is complex (i.e., composed of two or more elementary steps, as shown in Fig. 4.1b), then the observed rate constant k_{obs} is a composite of the individual rate constants for each of the elementary steps k_1, k_2, k_3, and so on, and E_a, derived from k_{obs}, is not a proper measure of any of the individual barrier heights. However, a complex reaction can be broken down into its individual steps, each with its own rate constant and individual activation energy.

Let us now consider the *thermodynamic* characterization of the transition state through a useful pseudothermodynamic approach termed *transition state theory*.

5.4 TRANSITION STATE THEORY

We noted previously (Section 4.1) that for a single elementary reaction step the reactant must pass through a critical structure, termed the transition state, which is the point of highest energy along the reaction profile (Fig. 4.1a). Transition state theory tackles the problem of characterizing the transition state by applying the tools of thermodynamics. Of course the transition state, being a transient point along the reaction profile, cannot formally be considered in thermodynamic terms. Thus the validity of transition state theory rests on a number of key assumptions.

The primary assumption of transition state theory is that for any elementary process the *transition state is in thermodynamic equilibrium with the reactants.* For the forward reaction this may be expressed as

$$A + B \rightleftharpoons [TS^{\ddagger}] \rightarrow \text{Products} \qquad (5.12)$$

where the equilibrium constant for transition state formation $K^{\ddagger}$, by analogy with a regular equilibrium constant, is given by

$$K^{\ddagger} = [TS^{\ddagger}]/[A][B] \qquad (5.13)$$

If the reaction is reversible, a similar relationship may be written for the reverse process, in which products are transformed into reactants.

A second important element of transition state theory that may be derived from statistical mechanics, is that the universal rate constant for decomposition of the transition state $k^{\ddagger}$ is given by

$$k^{\ddagger} = kT/h \qquad (5.14)$$

where h is Planck's constant, k is Boltzmann's constant, and T is the temperature.

Since the rate of the reaction in terms of reactant concentrations is given by

$$\text{Rate} = k[A][B] \qquad (5.15)$$

where k is the rate constant for the reaction, and, in terms of the transition state concentration, is given by

$$\text{Rate} = k^{\ddagger}[TS^{\ddagger}] \qquad (5.16)$$

Eqs. (5.15) and (5.16) may be combined to give

$$k = k^{\ddagger}[TS^{\ddagger}]/[A][B] \qquad (5.17)$$

If we substitute $K^{\ddagger}$ from Eq. (5.13) we may write

$$k = k^{\ddagger}K^{\ddagger} \qquad (5.18)$$

Substituting for $k^{\ddagger}$ from Eq. (5.14) and rearranging we obtain

$$K^{\ddagger} = kh/kT \qquad (5.19)$$

Thus in Eq. (5.19) we have a relationship between the equilibrium constant for transition state formation $K^{\ddagger}$ and the rate constant for the reaction k.

This central idea wherein the transition state is in equilibrium with reactants and products really must be treated as an assumption since, as we discussed in Chapter 4, the transition state is not a regular species, but lacks a vibrational degree of freedom that corresponds to the reaction coordinate. We make the assumption, however, since by treating the transition state as a species in thermodynamic equilibrium with reactants and products, *we are now able to apply the equations of chemical equilibrium to reaction rates.* Thus, by analogy with the equation relating the free energy of a reaction ΔG°, with the equilibrium constant K

$$\Delta G^{\circ} = -RT \ln K \qquad (5.20)$$

we may write

$$\Delta G^{\ddagger} = -RT \ln K^{\ddagger} \qquad (5.21)$$

where $\Delta G^{\ddagger}$ is the free energy of activation for the reaction (i.e., the free energy difference between the transition state and the reactants). We can now substitute the value of $K^{\ddagger}$ from Eq. (5.19) to obtain

$$\Delta G^{\ddagger} = -RT \ln (kh/kT) \qquad (5.22a)$$

or in exponential form

$$k = (kT/h) \exp (-\Delta G^{\ddagger}/RT) \qquad (5.22b)$$

Equations (5.22a) and (5.22b) are two forms of what is termed the Eyring equation. The Eyring equation provides *a direct relationship between the rate constant for an elementary reaction and its free energy of activation.* In other words transition state theory enables the rate constant for an elementary reaction to be related to the free energy of the transition state.

Just as we can define free energy of activation $\Delta G^{\ddagger}$, so we can define the terms, *enthalpy of activation* $\Delta H^{\ddagger}$ and *entropy of activation* $\Delta S^{\ddagger}$. Thus by analogy with the corresponding thermodynamic equation

$$\Delta G^{\circ} = \Delta H^{\circ} - T\Delta S^{\circ} \qquad (5.23)$$

we may write

$$\Delta G^{\ddagger} = \Delta H^{\ddagger} - T\Delta S^{\ddagger} \qquad (5.24)$$

Substituting for $\Delta G^{\ddagger}$ in Eq. (5.22a) gives

$$\Delta H^{\ddagger} - T\Delta S^{\ddagger} = -RT \ln (kh/kT) \qquad (5.25)$$

and rearranging in terms of the rate constant k, leads to an alternative form of the Eyring equation

$$k = (kT/h) \exp (-\Delta H^{\ddagger}/RT) \cdot \exp (\Delta S^{\ddagger}/R) \qquad (5.26)$$

in which k is related to the two activation parameters $\Delta H^{\ddagger}$ and $\Delta S^{\ddagger}$. Let us now determine how we can measure these two parameters, and then consider their mechanistic significance.

5.4.1 Evaluation of $\Delta H^{\ddagger}$ and $\Delta S^{\ddagger}$

The analogy between the Arrhenius equation [Eq. (5.5)] and the form of the Eyring equation shown in Eq. (5.26), may be used to relate the Arrhenius parameters E_a and A to the activation barriers $\Delta H^{\ddagger}$ and $\Delta S^{\ddagger}$, which are derived from transition state theory. This analogy is useful, since from the experimentally derived values of E_a and A one can obtain values for $\Delta H^{\ddagger}$ and $\Delta S^{\ddagger}$.

Expressing the Eyring equation [Eq. (5.22)] and the Arrhenius equation [Eq. (5.5)] in logarithmic form gives

$$\ln k = \ln (k/h) + \ln T - \Delta H^{\ddagger}/RT + \Delta S^{\ddagger}/R \qquad (5.27)$$

and

$$\ln k = \ln A - E_a/RT \qquad (5.28)$$

Differentiating Eqs. (5.27) and (5.28) with respect to $1/T$ and equating the two differential expressions gives

$$d (\ln k)/d(1/T) = -T - \Delta H^{\ddagger}/R = -E_a/R \qquad (5.29)$$

Rearranging Eq. (5.29) gives

$$E_a = \Delta H^{\ddagger} + RT \qquad (5.30)$$

Thus in Eq. (5.30) we have a simple means of evaluating $\Delta H^{\ddagger}$ from the experimentally determined E_a value. In practice, however, the difference between E_a and $\Delta H^{\ddagger}$ is quite small, being just 0.6 kcal mol^{-1} at room temperature.

The magnitude of $\Delta S^{\ddagger}$ may be obtained from the empirically measured value of A. The relationship between the two is obtained by substituting Eq. (5.30) into the Arrhenius equation [Eq. (5.5)], and comparing the result with Eq.

(5.26). This leads to

$$A = (ekT/h) \exp (\Delta S^{\ddagger}/R) \qquad (5.31)$$

So in Eqs. (5.30) and (5.31) we have the means to evaluate the thermodynamic parameters $\Delta H^{\ddagger}$ and $\Delta S^{\ddagger}$ from the experimentally derived parameters E_a and A.

Some values of these parameters, and their contribution to the rate constants, are provided in Table 5.1. These values provide some "feel" as to how the magnitude of the activation parameters $\Delta H^{\ddagger}$ and $\Delta S^{\ddagger}$ affect the magnitude of the rate constant calculated from Eq. (5.26). We see that reactions whose $\Delta H^{\ddagger}$ value is *less* than 10 kcal mol^{-1} contribute to the reaction being fast ($k > 2.85 \times 10^5$ without any $\Delta S^{\ddagger}$ contribution), while reactions whose barriers are greater than 25 kcal mol^{-1} contribute to the reaction being slow ($k < 2.80 \times 10^{-6}$ without any $\Delta S^{\ddagger}$ contribution). From Table 5.1 we also see that $\Delta S^{\ddagger}$ contributions to the rate constant can be significant and may speed up, or slow down, the "enthalpic" rate by factors as large as 10^7. Positive entropies of activation enhance the rate while negative entropies lower the rate.

5.4.2 How Fast Can Reactions Be?

The Eyring equation may be used to estimate the *maximum* rate constant for a chemical reaction. If we take a unimolecular reaction that has a zero free energy of activation, the exponential part of the equation equals 1, and the rate constant for the reaction becomes kT/h, which equals 6×10^{12} s^{-1} at room temperature.

TABLE 5.1. Effect of $\Delta H^{\ddagger}$ and $\Delta S^{\ddagger}$ Values on the Magnitude of the Rate Constant k, based on the Eyring Equation [Eq. (5.26)]^a

$\Delta H^{\ddagger}$ (kcal mol^{-1})^b	$(kT/h) \exp(-\Delta H^{\ddagger}/RT)$ (s^{-1})	$\Delta S^{\ddagger}$ (cal mol^{-1} deg^{-1})^c	$\exp(\Delta S^{\ddagger}/R)$
1 (4.2)	1.15×10^{12}	30 (126)	3.63×10^6
2 (8.4)	2.12×10^{11}	25 (105)	2.93×10^5
5 (21)	1.33×10^9	20 (84)	2.36×10^4
10 (42)	2.85×10^5	15 (63)	1.91×10^3
15 (63)	6.10×10^1	10 (42)	1.54×10^2
20 (84)	1.31×10^{-2}	5 (21)	2.14×10^1
25 (105)	2.80×10^{-6}	0	1
30 (126)	6.00×10^{-10}	-5 (-21)	8.07×10^{-2}
40 (167)	2.76×10^{-17}	-10 (-42)	6.50×10^{-3}
		-15 (-63)	5.25×10^{-4}
		-20 (-84)	4.23×10^{-5}
		-30 (-126)	2.75×10^{-7}

^{a}Adapted from Table 2.17 in Ref. 3.
^{b}Values in parentheses in kilojoules per mole (kJ mol^{-1}).
^{c}Values in parentheses in joule per mole per degree (J mol^{-1} deg^{-1}).

It is noteworthy that this rate constant approaches the vibrational frequency of a bond, which is of the order of 10^{13}–10^{14} s^{-1}. Clearly, this last value must represent the upper limit for any unimolecular rate constant, since the time required for any bond-breaking process must be *at least* that of a single vibration.

For bimolecular reactions the situation is more complicated; before any reaction can take place the reacting molecules must come into close proximity with one another. For such a reaction we can separate the approach of the reacting molecules and their actual reaction into two discrete steps. Accordingly, the kinetic scheme may be written as

$$A + B \underset{k_{-d}}{\overset{k_d}{\rightleftharpoons}} [\,A\text{---}B\,] \xrightarrow{k_1} \text{Products} \qquad (5.32)$$

where k_d is the rate constant for diffusion of the reactants to form the encounter complex, k_{-d} is the rate constant for complex break up into separated reactants, and k_1 is the rate constant for the actual reaction. If we assume the steady-state approximation for complex formation, we may write

$$k_{obs} = k_d k_1 / (k_{-d} + k_1) \qquad (5.33)$$

Thus the observed rate constant k_{obs} for the overall reaction of A and B to give products may be written in terms of the individual rate constants.

If all complexes once formed react rapidly to form products (i.e., $k_1 \gg k_{-d}$), then k_{obs} will just become k_d. In other words, such a reaction will be *diffusion controlled* and independent of k_1. The observed rate constant will then be at its maximum. This rate is of course variable, and for reactions in solution, they depend on the nature of the reactant molecules and the solvent, but it normally lies in the range 10^9–10^{10} L mol^{-1} s^{-1}. So what is the maximum rate constant for a bimolecular reaction in solution? About 10^{10} L mol^{-1} s^{-1}. If, however, separation of the complex back into the constituent reactants competes favorably with the actual reaction then $k_{obs} < k_d$ and its precise value will depend on the magnitude of the fraction $k_1/(k_{-d} + k_1)$.

5.4.3 Mechanistic Significance of $\Delta H^{\ddagger}$

We pointed out that $\Delta H^{\ddagger}$ measures the enthalpic barrier that reactants need to overcome in order to be converted into products. The actual value provides some indication of the extent of bond breaking and bond making (if it takes place) in the transition state. For example, the value of E_a (and $\Delta H^{\ddagger}$) for the homolysis reaction of ethane

$$CH_3\text{---}CH_3 \rightarrow 2CH_3\cdot \qquad (5.34)$$

is about 86 kcal mol^{-1}, effectively the same as the C—C bond strength. This indicates that in the transition state the C—C bond is essentially broken.

Alkene isomerization provides another example of this type of application. The values of $\Delta H^{\ddagger}$ for the cis–trans isomerization reaction of different alkenes has been found to be closely related to their π-bond strengths. This finding suggests that the mechanism for isomerization involves rotation about the double bond, through a transition state in which the two alkylene groups are perpendicular to one another.

In general one can say that large values of $\Delta H^{\ddagger}$ are indicative of a transition state that has undergone substantial bond breaking (with little compensating bond making). Conversely, reactions with low values of $\Delta H^{\ddagger}$ are suggestive of a transition state in which there is little bond breaking.

A more quantitative means of assessing the significance of $\Delta H^{\ddagger}$ values is through the use of Benson group additivities. The approach involves the estimation of ΔH_{f}° for some species, suspected of being a reaction intermediate or transition state with the aid of Group tables.[4] Thus by adding the ΔH_{f}° values listed in the table for all the groups within the molecule, one can obtain a good estimate of the ΔH_{f}° for the structure in question. The experimental $\Delta H^{\ddagger}$ can then be compared with the estimated ΔH_{f}° values of reactants and postulated intermediates or transition states, to establish whether the ΔH_{f}° values for the postulated species are indeed consistent with the experimental $\Delta H^{\ddagger}$ value. For example, if, using the Benson additivity scheme, a particular intermediate is found to have a ΔH_{f}° value that is 30 kcal mol^{-1} *higher* than that of the reactants, then such an intermediate cannot be postulated for a reaction whose measured $\Delta H^{\ddagger}$ is just 15 kcal mol^{-1}.[5]

5.4.4 Mechanistic Significance of $\Delta S^{\ddagger}$

The magnitude of $\Delta S^{\ddagger}$ can also provide some mechanistic insight into the structure of the transition state. A negative value of $\Delta S^{\ddagger}$ means that the entropy of the transition state is *lower* than that of the reactants, and this signifies a transition state that is *more* ordered than the reactants. Similarly, a positive value of $\Delta S^{\ddagger}$ signifies a transition state that is *less* ordered than the reactants. For this reason bimolecular reactions, such as S_N2 and cycloaddition, show large *negative* values of $\Delta S^{\ddagger}$ (typically between -10 and -40 cal mol^{-1} deg^{-1}), since these reactions require two reacting molecules to approach one another in some precise orientation. The result is a highly ordered transition state with a low entropy of formation. The S_N1 reactions, by contrast, tend to have either positive or small negative $\Delta S^{\ddagger}$ values. Precise values are difficult to specify since they are highly solvent dependent.

One point that should be noted is that $\Delta S^{\ddagger}$ data for reactions that are second (or higher) order should be interpreted with caution; the numerical values of $\Delta S^{\ddagger}$ depend on the reference standard state, which in turn, is governed by the units of the rate constant (L mol^{-1} s^{-1}, cm^3 mol^{-1} s^{-1}, etc.). For this reason

$\Delta S^{\ddagger}$ values in such cases are probably best interpreted when used in a *comparative* rather than in an *absolute* way.

Unimolecular reactions can exhibit both positive and negative $\Delta S^{\ddagger}$ values. Bond homolysis reactions, such as that of di-*tert*-butyl peroxide

$$(CH_3)_3C-O-O-C(CH_3)_3 \rightarrow 2(CH_3)_3C-O\cdot$$

$$\Delta S^{\ddagger} = +10 \text{ cal mol}^{-1} \text{ deg}^{-1} \tag{5.35}$$

typically show positive $\Delta S^{\ddagger}$ values, since the transition state for the reaction (in which the O—O bond has been substantially broken) provides for *greater* structural freedom than in the reactants. On the other hand, the negative $\Delta S^{\ddagger}$ values found for the isomerization of *iso*-propenyl allyl ether to allyl acetone

$$\tag{5.36}$$

$$(\Delta S^{\ddagger} = -7.6 \text{ cal mol}^{-1} \text{ deg}^{-1})$$

and the electrocyclic ring closure of hexatriene

$$\tag{5.37}$$

$$(\Delta S^{\ddagger} = -7 \text{ cal mol}^{-1} \text{ deg}^{-1})$$

suggest highly structured cyclic transition states in which the conformational flexibility present in the reactant is partly lost. Thus the magnitude and sign of $\Delta S^{\ddagger}$ can provide some information regarding the *structure* of the transition state.[5]

5.5 HAMMOND POSTULATE

As already noted, understanding reactivity requires understanding the transition state. Unfortunately, the transition state, being a transient point on the reaction surface, is difficult to characterize. From transition state theory, we have seen how we can relate the *energy* of the transition state to the rate constant for the reaction. However, obtaining information regarding the *structure* of the transition state, even indirectly, is difficult. For this reason any general rules that will provide information on the structure of the transition state would be very useful.

An important postulate bearing on the question of transition state structure was proposed by Hammond:[6] *"If two states, as for example, a transition state and an unstable intermediate, occur consecutively during a reaction process and have nearly the same energy content, their interconversion will involve only a small reorganization of molecular structure.''*

Consider a highly exothermic and a highly endothermic reaction, as illustrated in Figs. 5.1a and b. Since the transition state for the endothermic reaction is likely to be similar in energy to the products, as indicated in Fig. 5.1a, Hammond's postulate suggests that the structure of the transition state will be product-like. This means that the amount of structural reorganization that takes place in the system on proceeding from reactants to transition state, will be a large proportion of the total structural reorganization that takes place during the reaction. Thus we can classify the transition state as *late*. By a similar argument we can deduce that for an exothermic reaction (Fig. 5.1b), the transition state is close in structure to the reactants, that is, the transition state is *early*.

The Hammond postulate refers to a specific situation (highly exothermic or endothermic reactions) so its application is necessarily a limited one. For example, it does not formally consider reactions that are not highly exothermic (or endothermic). This point is worth noting since, as we will discuss later, *extensions* of the Hammond postulate appear to be much less reliable than the Hammond postulate itself.

An important application of the Hammond postulate relates to the relative stability of reactive intermediates. Consider the formation of a series of carbocations R^+ in the ionization step of a family of S_N1 reactions with different R groups

$$R-X \rightarrow R^+ + X^- \tag{5.38}$$

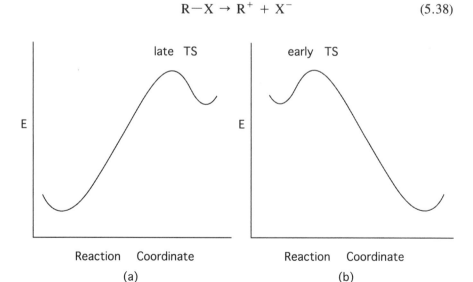

Fig. 5.1. Schematic energy diagram illustrating the Hammond postulate. (a) A highly endothermic reactions will have a product-like transition state. (b) A highly exothermic reaction will have a reactant-like transition state.

The relative stability of R^+ may be obtained by measuring the relative rate constant of the ionization reaction [Eq. (5.38)]. Carbocations that are formed rapidly are concluded to be more stable than those that are generated more slowly. Inspection of Fig. 5.2 shows this conclusion rests on the validity of the Hammond postulate.

The rate constant of ionization of $R-X$ is a direct measure of $\Delta G^{\ddagger}$, the free energy of activation that leads to the formation of the carbocation, and is *not* a measure of the free energy of the carbocation itself. However, since the carbocation is a high-energy intermediate, the Hammond postulate suggests that *the transition state leading to its formation will be similar in structure and energy to the cation itself.* Thus the rate of formation of a relatively stable cation R_1^+, will be *greater* than the rate of formation of a less stable cation R_2^+. The general implication of this kind of application is that *substituent effects on high-energy intermediates, and the transition states that lead to their formation, are likely to be closely related.* In other words, knowing the effect of substituents on the stability of reaction intermediates, also tells us the effect of substituents on the *rate of reaction* leading to formation of those intermediates. It is the basis for the general expectation that relatively stable intermediates will form more rapidly than less stable ones.

Over the years there have been various attempts to extend the ideas that are implicit in the Hammond postulate to the whole spectrum of reaction types—not just highly exothermic and endothermic reactions. These proposals led to the development of some of the familiar concepts in physical organic chemistry. The use of linear free energy relationship parameters as measures of transition

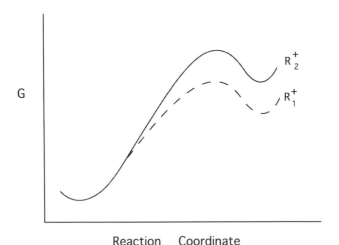

Reaction   Coordinate

Fig. 5.2. Schematic energy diagram illustrating the way the rate of formation of a high-energy intermediate is related to the energy of the intermediate. In the S_N1 dissociation of R_1-X and R_2-X, the more stable cation R_1^+ will be formed more rapidly than the less stable cation R_2^+.

state structure, the reactivity–selectivity principle, and so on, are obvious examples. However, it now appears that the generality of these ideas is more limited than the Hammond postulate itself, and with hindsight it seems that certain of these applications may have been overextended. Now let us critically consider the more important of these ideas as well as their possible limitations.

5.6 BELL–EVANS–POLANYI PRINCIPLE

Over 50 years ago some fundamental concepts of physical organic chemistry were laid down when Bell, Evans, and Polanyi attempted to elucidate the factors governing chemical reactivity by considering general substitution reactions of the type

$$A + B{-}C \rightarrow A{-}B + C \tag{5.39}$$

They recognized that the barrier to the reaction derives from the need to stretch the B—C bond and overcome initial A---B repulsion before the energetic benefits of A—B formation can be obtained. By drawing energy curves for A—B and B—C stretching one can qualitatively derive the widely observed correlation between rates and equilibria and make some predictions regarding the effect of reaction exothermicity on transition state structure. Let us see how this is done.

Consider the energy curves for B—C bond stretching and A—B bond formation superimposed on an energy diagram, as shown in Fig. 5.3. (Note that this superimposition is actually artificial, since the A—B and B—C axes are different. Strictly speaking, there is no formal meaning to the superimposition of these two curves.) If the barrier to substitution $\Delta E^{\ddagger}$ is approximated by the energy difference between reactants and the crossing point of the two curves, it becomes apparent that the barrier for the concerted substitution reaction, in which A displaces C, will be *lower* than that for a two-step pathway in which B—C initially breaks up into B and C (whose energy is defined by the point A + B + C in Fig. 5.3), followed by the combination of A and B. In other words the ability to partially form the A—B bond before the B—C bond is fully broken, leads to a reduction in the activation barrier of the concerted process in comparison to the two-step pathway.

Now consider two related processes

$$A_1 + B{-}C \rightarrow A_1{-}B + C \tag{5.40a}$$

$$A_2 + B{-}C \rightarrow A_2{-}B + C \tag{5.40b}$$

where the A_1—B bond is weaker than the A_2—B bond. A Bell–Evans–Polanyi diagram for these two processes is illustrated in Fig. 5.4. It is apparent that

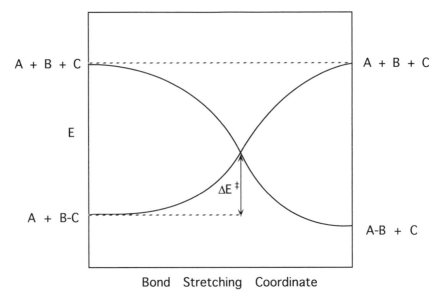

Fig. 5.3. Bell–Evans–Polanyi diagram showing barrier formation in the reaction A + B—C → A—B + C, through the superimposition of energy curves for B—C bond stretching and A—B bond formation.

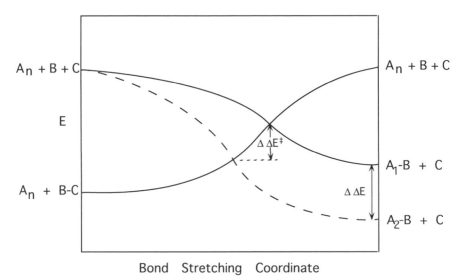

Fig. 5.4. Bell–Evans–Polanyi diagram showing the effect of the A—B bond strength on the relative barrier height and location of the transition state in the reaction A + B—C → A—B + C. Strengthening the A—B bond leads to a lower barrier and an earlier transition state.

the crossing point for the A_2-B curve is *lower* and *earlier* than that for the A_1-B curve. Two general conclusions immediately follow:

1. *Product Stabilization Induces Stabilization of the Transition State (and therefore to a Reduction in the Activation Barrier).* Thus there is a relationship between *changes in the reaction energy* ΔE, and *changes in the reaction barrier* $\Delta E^{\ddagger}$ given by

$$\Delta\Delta E^{\ddagger} = \alpha\Delta\Delta E \qquad (5.41)$$

From simple geometric considerations (Fig. 5.4), it is apparent that $\Delta\Delta E^{\ddagger}$ is just some fraction of $\Delta\Delta E$ (assuming that the A_1-B $-$ A_2-B energy difference decreases monotonically as these two bonds are stretched), so that α lies in the range 0 to 1. The energy relationship of Eq. (5.41) is the early theoretical justification for the rate-equilibrium relationship—the observation that for a series of related reactions, rates and equilibria often (but not always!) correlate with one another. Since rates and equilibria are defined in free energy terms G, rather than energy E, the appropriate free energy expression that describes the rate-equilibrium relationship is then

$$\Delta\Delta G^{\ddagger} = \alpha\Delta\Delta G^{\circ} \qquad (5.42)$$

where $\Delta\Delta G^{\ddagger}$ is the difference in the free energy of activation for the two reactions, and $\Delta\Delta G^{\circ}$ is the difference in the free energies of the two reactions.

2. *A More Exothermic Reaction Will Have an Earlier Transition State than a Less Exothermic Reaction.* From Fig. 5.4 it can be seen that formation of the *more* stable A_2-B product leads to a crossing point in which there is less $B-C$ stretching, and *less* $A-B$ bond making. One may view this conclusion as a *differential extension of the Hammond postulate* (though historically Bell, Evans, and Polanyi preceded Hammond), in that it makes a statement regarding the structural difference between two transition states for two related reactions that differ in their exothermicity.

The above rules enable the effect of substituents on reactivity to be assessed. Consider the general S_N2 reaction

$$N^- + R-X \rightarrow N-R + X^- \qquad (5.43)$$

We might ask the following questions: How will substituent effects on N and X affect the rate of reaction? How will the structure of the transition state vary as a result of these perturbations?

Substituting the leaving group X by an electron-withdrawing group (such as $-NO_2$) stabilizes the product X^- (by dispersing the negative charge). According to the above rules, product stabilization will bring about a *faster* reaction, whose transition state structure will be *earlier*. The same substituent on the nucleophile will lead to the opposite result: a stabilized reactant, a *slower* reaction, and a *later* transition state.

A major problem is that the two rules described above do not always lead to the correct prediction; breakdowns in the rate-equilibrium relationship are common. Subsequently, we will see (Section 6.5) that this failure can be attributed to the simplistic nature of the assumptions inherent in the Bell–Evans–Polanyi model. Unfortunately, the Bell–Evans–Polanyi approach provides no indication as to when failures of the model are likely to occur, or how these might be remedied.

5.7 POTENTIAL ENERGY SURFACE MODEL

The Bell–Evans–Polanyi model deals with substituent effects *along the reaction coordinate*. In other words the model predicts the manner in which substituent effects on *reactants* or *products*, will affect the energy and structure of the transition state. However, an important question is how substituent effects *on structures that do not lie along the reaction coordinate* (e.g., potential intermediates) will influence the structure and energy of the transition state and modify the reaction pathway. Let us once again consider the S_N2 reaction [Eq. (5.43)] to illustrate this point.

For the S_N2 reaction of N^- and $R—X$, one might ask the question How will a substituent that affects the energy of the carbocation R^+ influence the reactivity of $R—X$ and the structure of its transition state. We ask the question *even though the carbocation intermediate is not actually formed in this reaction*. A solution to this problem is provided by the potential energy surface (PES) model developed by Thornton,[7a] More O'Ferrall,[7b] and Jencks,[7c] which treats perturbations both *along* the reaction coordinate and *perpendicular* to the reaction coordinate. Hence, the PES model provides a more complete picture of reactions in which *two* bonds are being formed, or broken, during the reaction. Application of the model begins with a PES diagram of the kind discussed in Section 4.2. For a nucleophilic substitution reaction the appropriate PES diagram is shown in Fig. 5.5.

The diagram shows the reactants $N^- + R—X$ in the bottom-left-hand corner and the products $N—R + X^-$, in the top-right-hand corner. The vertical axis corresponds to changes in the $R—-X$ bond distance, and the horizontal axis to changes in the $N—-R$ bond distance. However, the surface also includes structures not lying along the reaction coordinate. This means that the top-left-hand corner represents the species $N^- + R^+ + X^-$ (i.e., $R—X$ bond breaking with no $N—R$ bond making), while the bottom-right-hand corner represents the $(N—R—X)^-$ pentacoordinate complex (i.e., $N—R$ bond making with no $R—X$ bond breaking). The energy axis is the one perpendicular to the plane of the paper and is not shown explicitly. Energy changes on the diagram are sometimes illustrated using contour lines (see Fig. 4.2a), though these lines are often omitted (as in Fig. 5.5). By omitting the energy contour lines a number of different reaction pathways for a particular reaction may be illustrated with the one PES diagram.

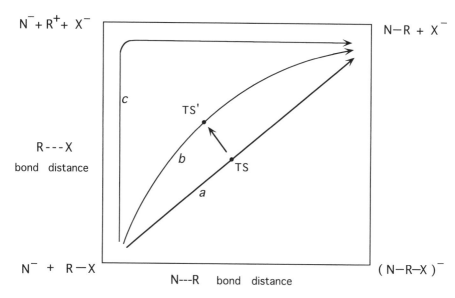

Fig. 5.5. Potential energy surface diagram for the general nucleophilic substitution reaction $N^- + R{-}X \rightarrow N{-}R + X^-$, showing mechanistic pathways for an S_N2 pathway through a "tight" transition state labeled *a*, through a "loose" transition state labeled *b*, and via an S_N1 pathway, labeled *c*.

Once the corners of the surface are labeled, we may proceed to specify different reaction pathways. An S_N2 pathway would be indicated by the arrow labeled *a*, which cuts across the surface in a diagonal fashion. This diagonal pathway specifies that $N{-}R$ bond making and $R{-}X$ bond breaking are taking place in a concerted fashion, as required by the S_N2 mechanism. An S_N1 pathway would follow the route indicated by the arrow labeled *c*. Initially, the $R{-}X$ bond is broken to give the carbocation R^+ (located in the top left-hand corner of the diagram) and then $N{-}R$ bond making takes place to yield the reaction products. Of course any number of additional reaction pathways on this surface are possible in principle. One such pathway is illustrated by the arrow labeled *b*, where the extent of $R{-}X$ bond-breaking is more advanced compared to $N{-}R$ bond-making than in the "tighter" S_N2 pathway *a*. Thus the precise location of each pathway on the reaction surface will depend on the extent to which $N{-}R$ bond-making and $R{-}X$ bond-breaking processes are synchronized.

How are such diagrams useful? The PES diagrams are of value because they allow us to visualize in a simple fashion the way substituent effects are likely to affect a particular reaction pathway. For example, let us say a particular system follows the S_N2 pathway, labeled *a*. It would be useful to know what effect a substituent change in that system will have on the reaction pathway. And how the structure and energy of the transition state will be affected?

The rules governing substituent effects on transition state structure and energy may be summarized as follows:

1. A substituent effect that lowers the energy of a species *on* the reaction coordinate (i.e., that lowers the energy of either reactant or product corners), will *stabilize* the transition state and move it structurally in the direction *away from* the corner that is stabilized. Because this effect relates to species on the reaction coordinate, it is termed a *Hammond effect*.

2. A substituent effect that stabilizes the energy of a species *perpendicular* to the reaction coordinate (i.e., that lowers the energy of an intermediate corner) will *stabilize* the transition state and move it structurally in the direction *toward* the corner that is stabilized. Because the response here is the opposite to that observed *on* the reaction coordinate, it is termed an *anti-Hammond effect*.

Let us now apply the PES model to the S_N2 reaction [Eq. (5.43)] by considering what effect a phenyl substituent on R will have on the S_N2 pathway (labeled *a*) and the structure of its transition state (marked by the point TS in Fig. 5.5). Since the main electronic effect of a phenyl group on the various structures at the corners of the PES diagram will be to stabilize R^+, any corner with R^+ in it will be affected. Accordingly, the top-left-hand corner of the diagram will be stabilized. Since this is a substituent effect that is *perpendicular* to the reaction pathway, according to Rule 2 its effect will be to *stabilize* the transition state and draw the transition state *toward* the R^+ corner. This substituent effect is shown by the small arrow leading to the new transition state position TS' and the new reaction pathway marked *b*. Thus incorporating the phenyl substituent into the R group is predicted to (a) enhance reactivity, (b) make the transition state *looser* (longer N--R and R--X distances), and (c) make the R group partially *carbocationic* in character.

If R^+ is stabilized even further (say by a *p*-methoxyphenyl group), then the perpendicular effect may draw the reaction pathway all the way into the carbocation corner. In such a case the reaction pathway *actually becomes a two-step process that proceeds via the R^+ intermediate* (pathway *c*). Thus the way in which two mechanisms, such as S_N1 and S_N2, relate to one another, and the mechanistic spectrum linking the two extremes, are presented in a particularly lucid fashion. With the two perpendicular corners of the diagram $(N-R-X)^-$ and $N^- + R^+ + X^-$, one can generate the full mechanistic spectrum of nucleophilic substitution, ranging from a tight S_N2 transition state at one extreme, through looser S_N2 transition states, in which some positive charge on the central carbon has developed, and leading finally to the two-step S_N1 mechanism at the other mechanistic extreme. This is the primary value of the PES model; in contrast to the one-dimensional reaction coordinate Bell–Evans–Polanyi diagrams, the PES model enables mechanistic changes that link concerted reactions and step-wise reactions (via a reactive intermediate) and the factors that control them, to be analyzed.

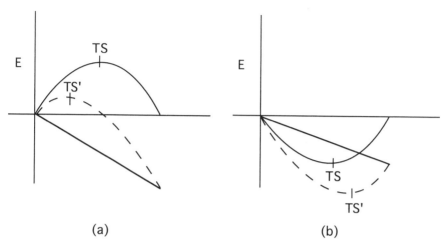

Fig. 5.6. Simple geometric model, based on the summation of a parabola and straight line, which provides a basis for understanding Hammond and anti-Hammond substituent effects on the structure and energy of a transition state. (a) Illustration of the Hammond effect and (b) illustration of the anti-Hammond effect (see text).

What is the basis for the two rules described above? A rationale for the rules may be obtained by using a simple geometric model first proposed by Thornton.[7a] This is shown in Fig. 5.6.

Along the reaction coordinate (Fig. 5.6a), the reaction profile (obtained by taking a cross section of the PE surface) may be approximated by an inverted parabola, with the transition state (labeled TS) located at the apex of the parabola. Stabilization of the product corner is induced by a linear perturbation, indicated by the straight line. Simple summation of the parabola and line functions (either geometrically or analytically) leads to a new parabola, represented by the dashed curve. It is easy to verify that the new transition state (marked TS′) is *lower* in energy and structurally *further* from the corner that was stabilized, in agreement with Rule 1.

The energy profile perpendicular to the reaction coordinate (Fig. 5.6b), (also obtained by taking a cross section through the PE surface), is approximated by a parabola, in which the transition state (labeled TS), is located at the parabola *minimum*. (Recall that in a *perpendicular* direction to the reaction coordinate, the transition state lies at a local minimum). Stabilization of a perpendicular corner is induced by a linear perturbation as before, and is again indicated by the straight line. In this case simple summation of the parabola and line functions also leads to a new parabola, represented by the dashed curve. However, it is apparent that in this case the new stabilized transition state (marked TS′) is *structurally closer to the corner that was stabilized*, as required by Rule 2.

Of course, a substituent may affect the energy of more than one corner in the PES diagram. Consider the effect of a substituent that stabilizes the nucleophile (e.g., an electron-withdrawing group, such as NO_2). This is illustrated in Fig. 5.7. Stabilizing the nucleophile will have the effect of stabilizing *two*

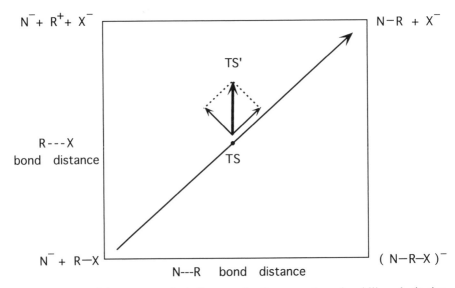

Fig. 5.7. Potential energy surface diagram for the general nucleophilic substitution reaction $N^- + R-X \rightarrow N-R + X^-$, showing the effect of a stabilizing substituent in the nucleophile on the location of the transition state. Total effect (moving from TS to TS') is the vector sum of Hammond and anti-Hammond contributions.

corners of the PES diagram, since the nucleophile in its reactant form N^- appears in both the top- and bottom-left-hand corners. Stabilizing the bottom-left-hand (reactant) corner will make the transition state more *product-like* (Rule 1), while stabilizing the top-left-hand (carbocationic) corner makes the transition state more *carbocationic* in character. The effect on the PES diagram of these two perturbations is shown in Fig. 5.7 by the two arrows, parallel and perpendicular to the reaction coordinate. The composite effect, obtained by vector addition of the two contributions, is illustrated by the arrow that leads to the new transition state, TS'. Thus the model predicts that stabilization of the nucleophile is likely to lead to *increased* $R-X$ bond breaking in the transition state, with only little change in the amount of $N-R$ bond making, since with regard to $N-R$ bond making, the two perturbations oppose one another. In summary, the PES model provides a simple means of evaluating the effect of substituents on reactions in which two bonds are made or broken, and in particular, to assess the influence of substituent effects on the perpendicular (or intermediate) corners.

How reliable are the predictions of the PES diagram? For the type of application discussed in the first example above (i.e., stabilization of a perpendicular corner), the PES approach has proved to be both reliable and very useful. The model correctly predicts the way in which stabilization of a potential *intermediate* will affect the reaction pathway of a concerted reaction. The application of the PES model to the spectrum of elimination reactions (E1 $-$ E2 $-$ E1$_{CB}$) is a particularly good example of this type of application. For applications in which two corners are stabilized at the same time, the evidence is less clear. The main problem is that unambiguous experimental procedures for assessing transition state structure in order to test the PES model are not readily available.

One difficulty that arises in the PES model is that charge and geometric progression along a reaction coordinate are considered to be related in a more or less linear fashion. For example, if we consider the S_N1–S_N2 surface illustrated in Fig. 5.5, it would appear that moving along the vertical R--X axis leads to concurrent R---X elongation and the buildup of ionic character $R^{\delta+}$ $X^{\delta-}$ in the bond. In fact, from the curve-crossing description of a transition state (Section 4.4), there is good reason to believe that bond extension and charge development are *not* related in a simple manner. The curve-crossing model teaches us that charge reorganization takes place predominantly in the avoided crossing region and therefore somewhat abruptly, rather than occurring in a gradual manner along the entire reaction coordinate. This being the case, some ambiguity arises as to whether the PES diagrams are intended to quantify the degree of *geometric progression* or *charge progression* in the transition state. Clearly, if charge and geometric progression are not related in a linear manner the two bond axes cannot simultaneously represent both parameters.

5.8 MARCUS THEORY OF ELECTRON TRANSFER

The context of what is commonly referred to as Marcus theory has undergone some transformation since the ideas were first proposed by Marcus in 1956.[8] Initially, what Marcus proposed was a quantitative physical model that predicts the rate of an electron-transfer reaction from a donor to an acceptor, typically, between two transition metal complexes. However, Marcus theory is now seen to contribute to an understanding of chemical reactivity that extends far beyond electron-transfer reactions. This is because some features of the theory carry over to chemical reactions in a general way. Thus, the theory provides a general framework for understanding the relationship between the kinetic barrier for a particular reaction and its thermodynamic driving force, that is, Marcus theory provides insight into the general relationship between rates and equilibria in chemical reactions. Furthermore, Marcus theory provides insight into the general problem of barrier formation in a chemical reaction. The important contributions of Marcus theory to an understanding of chemical reactivity cannot be overstated.

5.8.1 Qualitative Aspects

The qualitative elements of the Marcus theory of electron transfer may be demonstrated by considering the identity electron-transfer reaction

$$*Fe^{3+}(aq) + Fe^{2+}(aq) \rightarrow *Fe^{2+}(aq) + Fe^{3+}(aq) \qquad (5.44a)$$

where the * label is used to differentiate between the two Fe atoms. [It may seem odd that a discussion of the rate of electron transfer between metal ions appears in a text on organic reactivity. However, as was discussed in Chapter 4, the electron-transfer reaction is intimately related to the most basic organic reactions (e.g., nucleophilic substitution, electrophilic substitution, electrophilic addition, etc.) since these reactions also involve the shift of a single electron. So, the very same factors that govern the rates of electron-transfer reactions, also influence the rates of many basic organic processes. The relationship between polar and electron-transfer reactions for nucleophilic and electrophilic substitution reactions is described in Chapter 9.]

Why is there a barrier to the identity transfer of an electron from Fe^{2+} to Fe^{3+}? At first sight it might seem that all that transpires during that reaction is the transfer of an electron from the donor to the acceptor. However, this is not the case; nuclear reorganization invariably accompanies the electronic reorganization, and indeed, Marcus postulated that *it is the nuclear reorganization that is actually responsible for the barrier to the electron transfer.*

For electron transfer between metal ions, the nuclear reorganization that occurs involves the readjustment of the solvation sphere around the metal ions so as to accommodate the new charge on the ion. In the case of electron transfer between Fe^{2+} and Fe^{3+}

$$\left(^*Fe^{3+}\right) + \left(Fe^{2+}\right) \longrightarrow \left(^*Fe^{2+}\right) + \left(Fe^{3+}\right) \qquad (5.44b)$$

there is a tight solvation sphere around the triply charged Fe^{3+} ions (represented by a small circle) and a looser solvation sphere around the less highly charged Fe^{2+} (represented by a big circle). As a result of electron transfer, the solvation spheres need to reorganize. *So the reaction coordinate for this reaction is the change in the solvation spheres around the two metal ions.* In geometric terms this change corresponds to a change in the metal ion—ligand bond distance from a short Fe—solvent bond in the solvated Fe^{3+} ion to a longer Fe—solvent bond in the solvated Fe^{2+} ion. Such a reaction coordinate might seem less obvious than one for an S_N2 reaction, for example, since no obvious reaction trajectory is involved (like nucleophile—substrate approach in an S_N2 reaction) but it is a reaction coordinate, nonetheless.

Let us now generate the ground-state energy profile for the electron-transfer reaction from a plot of two energy curves: one representing the reactant electronic configuration R (that describes the initial electronic distribution) and the other by the product electronic configuration P (that describes the final electronic distribution). The plot of the two curves is illustrated in Fig. 5.8. The energy of the reactant curve R *increases* along the reaction coordinate. The reason: The solvation sphere of the Fe^{3+} ion expands to one that is *less* effective in solvating Fe^{3+}, while that for the Fe^{2+} ion contracts so it is also *less* effective in solvating the Fe^{2+} ion. Thus in the *product* geometry, R has an Fe^{3+} ion

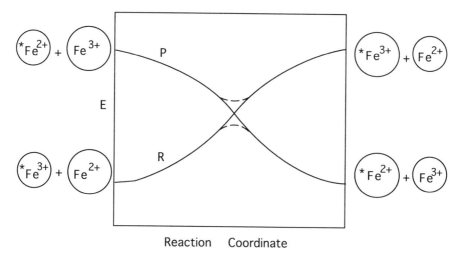

Fig. 5.8. Energy diagram showing the avoided crossing of reactant R and product P configuration curves for the identity reaction $*Fe^{3+} + Fe^{2+} \rightarrow *Fe^{2+} + Fe^{3+}$ in solution. Dashed lines indicate the region of avoided crossing due to strong mixing of R and P.

surrounded by a loose solvation sphere (whose geometry is suited to Fe^{2+}), and an Fe^{2+} ion surrounded by a *tight* solvation sphere (whose geometry is suited to an Fe^{3+} ion). Both Fe ions are poorly solvated so this is a high-energy state. In analogous fashion the product curve starts off high in energy, since it is in the *reactant* geometry where the two ions are poorly solvated. Then, as the system moves along the reaction coordinate, the solvation spheres change to the correct ones for the products.

We can now build up an energy profile for the electron-transfer process from the *avoided crossing* of the two configuration curves R and P. The avoided crossing is illustrated in Fig. 5.8 by the dashed lines in the crossing region. Thus interaction of the two configuration curves leads to the generation of two state curves: a *ground-state* curve and, above it, an *excited-state* curve. The two state curves (without the configuration curves) are shown explicitly in Fig. 5.9. It is the lower of these two curves that corresponds to the energy profile for a thermal electron-transfer reaction.

So how does the electron-transfer reaction between two metal ions actually take place? This question can be answered by stepping along the ground-state energy profile. Three stages are involved: (a) the coordination shell of the two ions reorganizes to some transient geometry (the transition state, labeled TS) at which an electron may transfer *without a change in energy*, (b) the actual electron transfer takes place, and (c) the coordination shell of the two ions relaxes from the transition state geometry to the equilibrium geometry of the products.

What will the solvation spheres look like when the electron transfer occurs?

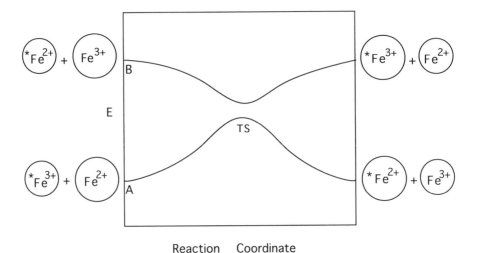

Reaction   Coordinate

Fig. 5.9. Energy diagram showing the avoided crossing of ground- and excited-state curves (obtained from the interaction of reactant and product configuration curves shown in Fig. 5.8) for the identity reaction $*Fe^{3+} + Fe^{2+} \rightarrow *Fe^{2+} + Fe^{3+}$ in solution.

Since the actual electron transfer only occurs when the energy of the system *before* electron transfer is the same as that *after* electron transfer, the solvation sphere around the two ions for the identity reaction of Eq. (5.44) must be *identical*. Thus the transition state for the reaction (labeled TS in Fig. 5.9), may be described by the resonance hybrid

$$TS = \left(*Fe^{3+} \right) + \left(Fe^{2+} \right) \longleftrightarrow \left(*Fe^{2+} \right) + \left(Fe^{3+} \right) \tag{5.45}$$

where the solvation spheres around the two ions are identical (intermediate between that for Fe^{2+} and Fe^{3+}), and the electron that is being transferred is delocalized between the two metal atoms. Thus the Marcus description reveals that the barrier to electron transfer arises from the need to reorganize the solvation shells of the two ions. This reorganization occurs in order "to pre-pare" the ions for the electron-transfer step.

Now the following question may arise. Why is an electron-transfer reaction not *initiated* by the electron transfer itself? Why do the solvation spheres distort *before* the actual electron-transfer phase? Why can the solvent reorganization not occur *after* the electron transfer? To answer these questions several points need to be made. First, *the transfer of an electron (or electron motion) takes place on a time scale considerably shorter than the time scale of nuclear motion*. This idea is termed the *Franck–Condon principle* and stems from the much lighter mass of the electron: Electrons travel much more rapidly than nuclei. The consequence of this principle is that when *an electron transfer results in a change in the energy of a system, the electron-transfer process is most likely to occur in a vertical manner* (i.e., *without any change in the geometry of the system*). The implications of this statement can be seen by reference to Fig. 5.10.

If an electron were to transfer from Fe^{2+} to Fe^{3+} *without* any prior solvent reorganization, this would mean a *vertical* excitation from point A in Fig. 5.10 (the reactants in their equilibrium solvation state) to point B. Point B represents the system *after* electron transfer *but with the solvation spheres unchanged*, that is, in a nonequilibrium solvation state. In effect, point B is an excited state of the reactants. From point B the system could then undergo solvation relaxation along the upper state energy curve, crossing over to the ground-state curve in the avoided crossing region, and then down to the products. This pathway is illustrated by the light arrows in Fig. 5.10. *Clearly such a reaction pathway would be of higher energy than the pathway along the ground-state reaction curve* (illustrated by the bold arrows in Fig. 5.10). So thermal electron transfer is initiated by solvation changes rather than by the electron transfer itself *because that pathway is lower in energy*. Actually, the high-energy pathway, which is initiated by electron transfer, is the *photochemical* pathway for the electron-transfer process, and it can be initiated by the appropriate photochemical excitation of the reactants from the ground surface (point A) to the excited-state surface (point B).

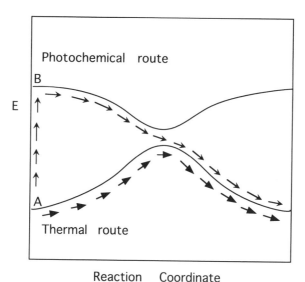

Fig. 5.10. Energy diagram showing the relationship between thermal and photochemical pathways for electron transfer. In the thermal pathway (bold arrows) geometric and solvation changes precede electron transfer, while in the photochemical pathway (light arrows) electron transfer precedes geometric and solvation changes.

In conclusion it is worth noting that the Marcus theory of electron transfer in its qualitative form, is the same as that of the curve-crossing model. In the terminology of the curve-crossing model, the barrier to electron transfer comes about through the avoided crossing of two configuration curves—the reactant curve labeled DA (D = donor, A = acceptor), and the product curve labeled $D^+ A^-$. Due to the poor overlap between the two configurations, the magnitude of the B term in the curve-crossing equation [Eq. (4.10)] is small (normally < 1 kcal mol^{-1}), but large enough to allow mixing of the configurations and the generation of an avoided crossing.

5.8.2 Quantitative Marcus Theory

The preceding qualitative description of electron transfer can be given a quantitative basis by evaluating the analytical form of the intersecting energy curves. If this is done the point of intersection, and hence the barrier to electron transfer, may be derived in a quantitative fashion. A particularly important point is the fact that though the equations were derived for electron-transfer reactions, subsequent examination of the reactivity patterns implied by the equations were found to apply to other chemical reactions as well. In fact, Marcus theory in its quantitative form provides a quantitative basis for the qualitative ideas associated with the Hammond postulate and the Bell–Evans–Polanyi principle. Specifically, it provides a *quantitative relationship between free energy of activation* $\Delta G^{\ddagger}$ *and free energy of reaction* ΔG°, *for a given reaction family*, or, put differently, the nature of the relationship between *reaction rates and reaction thermodynamics*.

At the heart of the theory is the Marcus equation

$$\Delta G^{\ddagger} = w + \Delta G_0^{\ddagger}(1 + \Delta G^{\circ}/4\Delta G_0^{\ddagger})^2 \qquad (5.46)$$

which relates the factors that contribute to an activation barrier $\Delta G^{\ddagger}$. The first of these, w, is just a work term that is electrostatic in nature, and describes the energy required to bring together the reactants in a preassociation step, and to separate the products, once formed. For cases where at least one of the reactants and one of the products are neutral, this term can be neglected, since electrostatic interactions are then relatively small. Thus the main part of the equation, and the one we will be concerned with, is the second term. According to this term the free energy of activation $\Delta G^{\ddagger}$, depends on two parameters—the *intrinsic barrier* $\Delta G_0^{\ddagger}$ and the *free energy of reaction* ΔG°. The first term, $\Delta G_0^{\ddagger}$, is defined as the free energy of activation *in the absence of any thermodynamic contribution*. This explains why it is termed the intrinsic barrier—it specifies the barrier height for the case in which there is no thermodynamic driving force, and is therefore just the value of $\Delta G^{\ddagger}$ when $\Delta G^{\circ} = 0$ [obtained by substituting $\Delta G^{\circ} = 0$ into Eq. (5.46) and assuming a zero work term]. Thus the Marcus equation breaks down the barrier of a chemical reaction into two separate contributions—a kinetic (or intrinsic) component that is independent of reaction thermodynamics, and a thermodynamic component that takes into account the contribution of the reaction thermodynamics. Inspection of Eq. (5.46) indicates that when the reaction is exergonic, (i.e., $\Delta G°$ is negative), $\Delta G^{\ddagger}$ is *smaller* than the intrinsic barrier while for endergonic reactions (i.e., $\Delta G°$ is positive) $\Delta G^{\ddagger}$ becomes *larger* than the intrinsic barrier, precisely as predicted by the Bell–Evans–Polanyi principle. So the usefulness of the Marcus equation is now apparent: *Knowing the intrinsic barrier $\Delta G_0^{\ddagger}$ for a particular reaction family, one can calculate the barrier height $\Delta G^{\ddagger}$ for each individual reaction from its free energy $\Delta G°$.*

Application of the Marcus equation would be particularly straightforward if reaction families really shared a common intrinsic barrier. In reality this is not generally the case. Let us return to our family of S_N2 reactions of N^- reacting with $R-X$ to demonstrate this point. Intrinsic barriers are most easily obtained by considering the barrier for an identity reaction. Thus for the S_N2 reaction of N^- and CH_3-X, the intrinsic barrier could be obtained from either of the exchange reactions

$$N^- + CH_3-N \rightarrow N-CH_3 + N^- \tag{5.47}$$

or

$$X^- + CH_3-X \rightarrow X-CH_3 + X^- \tag{5.48}$$

The problem is that the set of identity S_N2 reactions can have dramatically different barriers. For example, for $X = I$ the barrier in water is 22 kcal mol^{-1} while for $X = CN$ the barrier is 51 kcal mol^{-1}. A way around this complication is simply to average the two intrinsic barriers so that the intrinsic barrier for the reaction of I^- with CH_3-CN or CN^- with CH_3-I would be taken as $(22 + 51)/2 = 36.5$ kcal mol^{-1}. This value can then be utilized to calculate free energy barriers for both forward *and* back reactions

$$I^- + CH_3-CN \rightarrow I-CH_3 + CN^- \tag{5.49}$$

$$NC^- + CH_3-I \rightarrow NC-CH_3 + I^- \tag{5.50}$$

by using Eq. (5.46) into which one substitutes the appropriate value of $\Delta G°$.

The analytical form of the Marcus equation, depicted in Eq. (5.46), is rather cumbersome and additional insight into its essence may be obtained by use of a graphical representation of the equation. This is shown in Fig. 5.11.

In the graphical representation of Fig. 5.11, the barrier to any reaction is modeled by the intersection of two parabolas. The left-hand parabola represents the free energy curve of the reactant molecule [in the case of Eq. (5.50) this corresponds to $C-I$ stretching] while that of the right-hand parabola represents the free energy curve for the product molecule ($NC-C$ stretching in the above example). The free energy of the transition state is defined by the point of intersection of the two parabolas so that the barrier height is defined by the

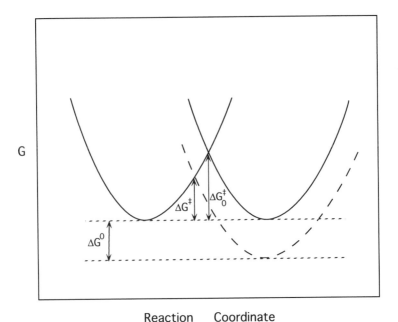

Fig. 5.11. Graphical representation of Marcus theory based on the intersection of two parabolas, and which defines the key terms—the intrinsic barrier $\Delta G_0^{\ddagger}$, the free energy of activation $\Delta G^{\ddagger}$, and the free energy of the reaction ΔG°.

distance between the left-hand parabola minimum and the parabolas intersection point. How is the model applied graphically?

Initially, the curvature of (or the distance between) the parabolas is set so that the barrier equals the intrinsic barrier $\Delta G_0^{\ddagger}$ when reactant and product free energy levels are the same. This is shown in Fig. 5.11 by the two parabolas, whose minima lie at the same energy level. The effect of reaction thermodynamics can now be assessed simply by raising or lowering the right-hand parabola to the appropriate free energy of reaction ΔG°. For example, lowering the parabola so as to model an exergonic reaction is depicted in Fig. 5.11 by the dashed parabola. The value of $\Delta G^{\ddagger}$ for the exergonic process is now defined by the *new* point of intersection. Of course raising the right-hand parabola would lead to an endergonic reaction with a *larger* value of $\Delta G^{\ddagger}$. We see therefore that the Marcus equation is just an analytical expression for a model of intersecting parabolas that specifies how changes in ΔG° affect the free energy of activation $\Delta G^{\ddagger}$.

The Marcus equation and its geometric expression provide information not only about barrier heights, but transition state geometries as well, by specifying whether the transition state is likely to be *early* or *late* along the reaction coordinate. From Fig. 5.11 it is apparent that the *more exergonic the reaction, the earlier the transition state.* Indeed the Marcus equation quantifies the position of the transition state (defined by the point of intersection of the parabolas) along the reaction coordinate. If the position of the transition state is defined by the parameter α, then from the Marcus equation we can write

$$\alpha = \frac{1}{2} + \frac{\Delta G°}{8\,\Delta G_0^{\ddagger}} \tag{5.51}$$

where α is defined in the range 0–1. Negative $\Delta G°$ values will lead to early transition states (i.e., $\alpha < \frac{1}{2}$), while positive $\Delta G°$ value will lead to late transition states (i.e., $\alpha > \frac{1}{2}$). So in Eq. (5.51) *we have a quantification of the second Bell–Evans–Polanyi rule.* Thus by knowing the exergonicity of a reaction and its intrinsic barrier, Eq. (5.51) suggests we can calculate the position of the transition state along the reaction coordinate. Some difficulties with the application of Marcus theory and the Marcus equation to organic reactions in general, will be discussed in Section 6.5.

5.9 REACTIVITY-SELECTIVITY PRINCIPLE

Chemical reactivity and selectivity are widely thought to be inversely related to one another, at least in a general way; reactive species are predicted to be relatively unselective in their chemical reactions, while stable species are predicted to be relatively selective. Reactions that conform to this general pattern can be found in many areas of chemistry. For example, in carbene chemistry, a highly reactive carbene, such as CH_2, is found to be less discriminate in its reaction with competing alkenes than the less reactive CF_2. Reactive radicals are also often found to be less selective than more stabilized radicals (examples are shown in Table 10.3). Consequently, this pattern of behavior has been generalized into what is termed the *reactivity–selectivity principle.*

Yet, despite the intuitive appeal of the principle and the many examples that offer experimental support for it, a critical evaluation of the reactivity–selectivity principle suggests that the number of cases where the principle is valid is not too different from the number of cases where it fails! For example, cations of widely varying reactivity have been found to exhibit constant selectivity toward competing nucleophiles. The wide-ranging failure in the case of cations has actually led to the establishment of a nucleophilicity scale: the N_+ scale. Moreover, while some radical reactions conform to the principle (see Table 10.3), others appear to violate it. For example, Hammett ρ values (see Chapter 6) for addition of 1-hexyl, cyclohexyl, and *tert*-butyl radicals to a family of substituted acrylic esters (for different substituents Z)

$$R\cdot + H_2C{=}C\overset{\displaystyle Z}{\underset{\displaystyle CO_2CH_3}{\big\langle}} \longrightarrow H_2C{-}\overset{\displaystyle Z}{\underset{\displaystyle R}{\overset{|}{C}}}\overset{}{\underset{\displaystyle CO_2CH_3}{\cdot\big\langle}} \tag{5.52}$$

R = 1-hexyl, cyclohexyl, or *tert*-butyl

increase from 2.4 to 3.1 to 4.1, respectively.[9] The ρ values constitute a relative measure of selectivity since a large ρ signifies that the rate constants for radical addition to different alkenes differ more than when the ρ value is small. Thus the selectivity of the radicals increases in the order

$$1\text{-hexyl} < \text{cyclohexyl} < \textit{tert}\text{-butyl}$$

However, radical reactivity increases in the same order. Thus we have a set of radicals in which the most reactive radical, *tert*-butyl, is also the most selective—a violation of the reactivity–selectivity principle.

So why is a principle that only appears to work one-half of the time formulated as a principle and engrained into chemical thinking? A key reason for this anomalous situation is that the inverted relationship between reactivity and selectivity can be derived from existing theoretical models of reactivity, such as the Bell–Evans–Polanyi model and Marcus theory. In other words, *on the basis of a simplified model of reactivity, an inverse relationship between reactivity and selectivity is anticipated*. Let us briefly demonstrate this using the Marcus equation.

From the Marcus equation [Eq. (5.46)], one can derive an expression for selectivity α by differentiating with respect to free energy. Thus,

$$\alpha = d(\Delta G^{\ddagger})/d(\Delta G^{\circ}) \tag{5.53}$$

α is a measure of selectivity because it specifies how sensitive the reaction barrier is to changes in reaction thermodynamics, or, expressed in rate terms, it measures the sensitivity of the rate constant to changes in the equilibrium constant; a small value of α signifies low selectivity, while a large value of α signifies high selectivity.

We have already discovered the parameter α in a different context in Eq. (5.51), where it provided a measure of transition state structure. Now we see that *the same expression provides a measure of selectivity as well*. Thus from the form of Eq. (5.51) we learn that as a reaction system is made exergonic (and more reactive), it becomes *less* selective ($\alpha < \frac{1}{2}$), while as the system is made endergonic (and less reactive), it becomes *more* selective ($\alpha > \frac{1}{2}$). In other words, according to a Marcus description of reactivity, *reactive systems with early transition states are predicted to be less selective than unreactive ones with late transition states*.

The problem with this description is that it is oversimplified. First, many potentially perturbing influences are ignored—solvent effects, complexation phenomena, mechanistic changeovers, and so on. However, even at a fundamental level the conclusions need to be questioned. The Marcus treatment, from which the reactivity–selectivity principle can be derived, is based on a simple two-curve description of reactivity. For electron-transfer reactions, this two-curve description (the crossing of DA and $D^{+}A^{-}$ curves) has proven to be excellent; there is normally no intermediate or potential intermediate that lies along an electron-transfer reaction coordinate.

For most organic reactions, however, a two-curve description is inadequate. For many organic reactions the transition state takes on electronic character that is *absent* in reactants and products. For example, the central carbon in an S_N2 reaction may take on either positive or negative charge, depending on its substitution pattern (see Chapter 9). Radical reactions often display charge-transfer character in the transition state (the so-called polar character; see Chapter 10). Concerted E2 elimination reactions often exhibit either carbocationic or carbanionic character in their transition states, and so on. In other words *some additional configuration that mixes into the transition state description is the norm for most organic reactions.* Once this additional configuration is present, the apparent link between reactivity and selectivity is easily broken; substituent effects on the energy of this intermediate configuration cause changes in both reactivity and selectivity that are not necessarily related in the anticipated fashion.

This is precisely the reason for the breakdown in the reactivity–selectivity principle for radical addition reactions [Eq. (5.52)]. The ionization potentials of alkyl radicals (see Table 7.3) decrease in the order primary > secondary > tertiary, reflecting the stabilities of the respective cations. Consequently, the donor ability of the three radicals *increases* in that order: 1-hexyl < cyclohexyl < *tert*-butyl. Thus, the *tert*-butyl radical, being a good donor radical (compared with cyclohexyl or 1-hexyl radicals) is more reactive since more of the charge-transfer configuration mixes into the transition state leading to greater transition state stabilization (for a configuration description of radical addition reactions, see Chapter 10). However, this greater charge-transfer from the radical into the alkene also results in a greater substituent effect on rates in the alkene; that is, larger ρ values. Consequently, the ρ value for *tert*-butyl addition (where there is more charge transfer) is larger than for 1-hexyl addition (where there is less charge transfer). *Thus, due to the role of the charge-transfer configuration in governing both radical reactivity and selectivity, the most reactive radical is also the most selective.* The corollary is clear: In real chemical situations the anticipated relationship between reactivity and selectivity cannot be relied upon to manifest itself.

The situation may be summarized as follows: While the reactivity–selectivity principle may be of theoretical interest and present in chemical systems as an *underlying* chemical phenomenon, it should not be utilized as a predictive tool. In real chemical systems the occurrence of factors that break the linkage between reactivity and selectivity are too frequent to allow its application as a general chemical principle.

REFERENCES

1. J. F. Bunnett. In *Investigation of Rates and Mechanisms of Reactions, Part I;* C. F. Bernasconi, Ed., Wiley, New York, 1986, pp. 253, 361.
2. For a detailed description of the role of kinetics in the study of reaction mechanism, see J. W. Moore and R. G. Pearson, *Kinetics and Mechanism*, 3rd ed. Wiley, New

York, 1981; K. J. Laidler, *Chemical Kinetics*, 2nd ed., McGraw-Hill Book Co., New York, 1965; F. Wilkinson, *Chemical Kinetics and Reaction Mechanisms*, Van Nostrand-Reinhold, New York, 1980.

3. T. H. Lowry and K. S. Richardson, *Mechanism and Theory in Organic Chemistry*, 3rd ed. Harper & Row, New York, 1987.

4. Tables of Benson group contributions, ΔH_f° and S° may be found in S. W. Benson, *Thermochemical Kinetics*, 2nd ed., Wiley-Interscience, New York, 1976, pp. 272–284.

5. A detailed discussion on the interpretation of activation parameters may be found in B. K. Carpenter, *Determination of Organic Reaction Mechanisms*, Wiley, New York, 1984.

6. G. S. Hammond, *J. Am. Chem. Soc.* **77**, 334 (1955).

7. (a) E. R. Thornton, *J. Am. Chem. Soc.* **89**, 2915 (1967); (b) R. A. More O'Ferrall, *J. Chem. Soc. B*, 274 (1970); (c) W. P. Jencks, *Chem. Rev.* **72**, 705 (1972).

8. R. A. Marcus, *J. Chem. Phys.* **24**, 966 (1956); **39**, 1734 (1963); R. A. Marcus, *Annu. Rev. Phys. Chem.* **15**, 155 (1964). For a recent comprehensive discussion of Marcus theory and the role of electron-transfer reactions within organic chemistry, see L. Eberson, *Electron Transfer Reactions in Organic Chemistry*, Springer-Verlag, Berlin, 1987.

9. B. Giese, *Angew. Chem. Int. Ed. Engl.* **22**, 753 (1983).

6

STRUCTURAL EFFECTS ON REACTIVITY

6.1 LINEAR FREE ENERGY RELATIONSHIPS

The magnitude of substituent (and solvent) effects on reactivity varies enormously—from very small to many orders of magnitude. This is no real surprise. However, an important observation is that there is very often an empirical relationship between the way substituent (or solvent) perturbations affect the rates and equilibria of *different* reactions. *In other words measuring the effect of a perturbation on one chemical reaction may lead to predictable changes when that perturbation is applied to a range of other reactions.* The fact that substituent and solvent effects for different systems are in some way interrelated has formed the basis for much of the research in physical organic chemistry over the past 30–40 years. As a class these relationships are termed *linear free energy relationships* (*LFERs*). We will explain the reason for the name later.

The importance of linear free energy relationships is easy to appreciate; they provide the chemist with useful information both of a qualitative and quantitative kind. Qualitatively these relationships enable us to make statements like "the nitro group is electron withdrawing" without having to know the identity of the substrate to which the group is attached. The nitro group would be expected to be electron withdrawing *regardless* of the substrate to which it is attached, and this particular property may then be used to make qualitative predictions about its effect on the reactivity of *any* substrate. Thus we would expect a nitro substituent in the aromatic ring to increase the acidity of benzoic acid, naphthoic acid, phenol, or toluene. However, the true power of LFERs lies at a *quantitative* level; LFERs provide a means of predicting unknown empirical data and are a most valuable source of mechanistic information. Let us now consider these relationships in some detail.

6.2 THE HAMMETT EQUATION

6.2.1 σ and ρ Parameters

The most important of the linear free energy relationships is the Hammett equation. In 1937 Hammett found that the effect of meta and para substituents on the acidity of a family of substituted benzoic acids correlated with their effect on the acidity of other acids. For example, if log K_a for a series of substituted benzoic acids is plotted against log K_a for a series of substituted phenylacetic acids, a good correlation (illustrated in Fig. 6.1) is obtained. This correlation tells us that the effect of substituents in one reaction family carries over quite well to another family. The slope of the line for the set shown (0.51), being less than 1, indicates that the family of phenylacetic acids is *less* sensitive to substituent effects than the family of benzoic acids. This lower sensitivity in the phenylacetic acids is readily explained since the aromatic ring is insulated from the reaction site by an additional methylene group. Clearly, as the distance of the substituent from the reaction site increases, substituent effects on reactivity are expected to become smaller.

The extraordinary thing is that this correlation of substituent effects applies not just to the ionization of structurally related acids, but covers a wide range of unrelated reactions as well. Recognizing this general feature, Hammett took the argument an important step further. *Hammett quantified the effect of substituents on any reaction by defining an empirical electronic substituent pa-*

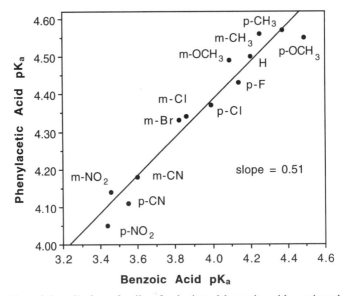

Fig. 6.1. Plot of the pK_a for a family of substituted benzoic acids against the pK_a for the corresponding family of substituted phenylacetic acids. The slope of the correlation line is 0.51.

rameter σ, which is derived from the acidity constants of substituted benzoic acids. Thus the electronic substituent parameter σ_X for any substituent X is defined by

$$\sigma_X = \log \frac{K_X}{K_H} \qquad (6.1)$$

where K_X and K_H are the acidity constants for the substituted benzoic acid and benzoic acid, respectively, in water at 25°C. Substituents that *increase* the acidity of the benzoic acid have *positive* σ values, while those that *decrease* the acidity have *negative* σ values. Each substituent gives rise to σ_{meta} and σ_{para} parameters, corresponding to substitution at the meta or para position of the benzene ring. Values for σ_{meta} and σ_{para} for some common substituents are listed in Table 6.1. (Ortho substituents are not included in such a table since these do not correlate well due to steric effects. It follows that steric effects, in contrast to electronic effects, are much less systematic and vary in a less predictable fashion. Therefore these effects cannot be treated in this way.)

What do the σ values signify? *The σ value of a substituent is a measure of the electron-withdrawing or electron-releasing ability of that substituent com-*

TABLE 6.1. Value of Electronic Substituent Constants[a]

Substituent	σ_{meta}	σ_{para}	σ^+	σ^-	σ_I
$CH{=}CH_2$	0.06	−0.04			0.11
$C{\equiv}CH$	0.29	0.25			0.29
$COCH_3$	0.38	0.50		0.87	0.30
CF_3	0.46	0.51			0.4
CH_2Cl	0.15	0.09	−0.01		0.17
CH_3	−0.06	−0.17	−0.31		0.0
CH_2CH_3	−0.05	−0.15	−0.30		0.0
CH_2OH	0.10	0.04			0.11
$COOH$	0.36	0.41			0.30
C_6H_5	0.09	0.01	−0.17		0.12
NH_2	−0.21	−0.63	−1.3		0.17
$NHAc$	0.16	−0.07	0.0		0.28
OCH_3	0.11	−0.28	−0.78		0.30
F	0.34	0.06	−0.07		0.54
Cl	0.37	0.22	0.11		0.47
Br	0.34	0.22	0.15		0.47
I	0.35	0.24	0.13		0.40
CN	0.61	0.65		0.90	0.57
NO_2	0.74	0.77		1.24	0.67
COO^-	−0.09	0.04	−0.03		−0.19
$N(CH_3)_3^+$	1.04	0.97			1.07
$S(CH_3)_2^+$	1.11	1.14			0.90

[a]Data taken from M. Charton, *Prog. Phys. Org. Chem.* **13**, 119 (1981), and from C. D. Ritchie and W. F. Sager, *Prog. Phys. Org. Chem.* **2**, 323 (1964).

pared with H. The σ value provides this information because the ionization of benzoic acid involves the deprotonation of a neutral molecule to form a negatively charged ion; deprotonation is *facilitated* by electron-withdrawing substituents that partially disperse the negative charge, but is *inhibited* by electron-releasing substituents that impede the charge dispersal. Thus substituents that are electron withdrawing have *positive* σ values, while those that are electron releasing have *negative* σ values. The larger the absolute value of σ, the larger the electronic effect. Since all substituent effects are obtained by reference to H, the σ value for H is (by definition) precisely 0. So in establishing the set of σ values, Hammett generated an empirical measure of electronic substituent effects.

With a set of substituent constants now established (defined on the basis of benzoic acid ionization constants), it becomes possible to assess substituent effects on the rate or equilibrium constants for other reactions. This is done by plotting the set of rate (or equilibrium) constants for a given reaction against the σ values of the substituents. The resulting correlations are expressed by

$$\log \frac{k_X}{k_H} = \sigma\rho \qquad (6.2a)$$

for rate constants, or

$$\log \frac{K_X}{K_H} = \sigma\rho \qquad (6.2b)$$

for equilibrium constants. These two equations are together known as the Hammett equation. The parameter ρ is the *slope* of the correlation line and measures the *sensitivity* of the particular reaction rate constant (or equilibrium constant) to substituent changes. Note that from the definition of σ, the ρ value for benzoic acid ionization must be 1. Thus reactions that have ρ values between 0 and 1 signify reactions of *lower* sensitivity to substituent effects than benzoic acid dissociation, while reactions that show ρ values greater than 1 signify *larger* sensitivity to substituent effects than benzoic acid ionization.

The sign of ρ is also informative. A *positive* ρ value indicates a reaction that is favored by electron-withdrawing substituents, while a *negative* ρ value indicates the reverse. We will discuss this point in greater detail later. Examples of reactions that obey the Hammett equation for rate and equilibrium processes, together with their measured ρ values, are listed in Table 6.2.

While the first eight entries in Table 6.2 demonstrate that different reaction types may correlate well with the Hammett σ parameter, for some reactions a good correlation is not obtained using this procedure. What this means is that substituent effects on benzoic acid ionization *cannot serve as a universal model of substituent effects in all chemical reactions.* Consequently, in cases where the σ scale is not applicable, an *alternative* scale based on an *alternative* model reaction needs to be devised. Let us now discuss the reasons for the limitations

TABLE 6.2. ρ Values for Rate and Equilibrium Constants of Selected Reactions

Reaction	Solvent	T(°C)	σ-Type	ρ
$ArCOOH \rightleftharpoons ArCOO^- + H^+$	H_2O	25	σ	1.00
$ArCOOH \rightleftharpoons ArCOO^- + H^+$	CH_3OH	25	σ	1.76
$ArCH_2COOH \rightleftharpoons ArCH_2COO^- + H^+$	H_2O	25	σ	0.56
$ArCOOCH_3 + HO^- \rightarrow ArCOO^- + CH_3OH$	Acetone (60%)	25	σ	2.23
$ArCHO + HCN \rightarrow ArCH(CN)OH$	Ethanol (95%)	20	σ	2.33
$ArCOCl + H_2O \rightarrow ArCOOH + HCl$	Acetone (95%)	25	σ	1.8
$ArSH + R\cdot \rightarrow ArS\cdot + RH$	$PhCH_3$		σ	-0.4
$ArO^- + EtI \rightarrow ArOEt + I^-$	CH_3CH_2OH	43	σ	-0.99
$Ar_3COH + H^+ \rightleftharpoons Ar_3C^+ + H_2O$	H_2O	25	σ^+	-3.6
$Ar_2CHCl + EtOH \rightarrow Ar_2CHOEt + HCl$	CH_3CH_2OH	25	σ^+	-5.1
$ArC(CH_3)_2Cl + EtOH \rightarrow ArC(CH_3)_2OEt$	CH_3CH_2OH	25	σ^+	-4.7
$ArOH \rightleftharpoons ArO^- + H^+$	H_2O	25	σ^-	2.1
$ArCl^a + CH_3O^- \rightarrow ArOCH_3 + Cl^-$	CH_3OH	25	σ^-	3.9

[a] Ar = 4-substituted-2-nitrophenyl.

of the σ scale and consider the way additional scales were set up to overcome these limitations.

When a substituted benzene ring is connected to some reaction center, the substituent transmits its electronic effect to the reaction center by two main mechanisms: the *inductive/field* effect (also termed the *polar* effect) and the *resonance* effect. The inductive/field effect involves the interaction of charges and dipoles in the substituent with the reaction center, and these appear to be transmitted partly through space, and partly through bonds, though the relative importance of these two modes of transmission remains controversial. We will consider these as one mechanism. (The existing terminology also reflects the controversial status of the mechanism; the term *inductive* emphasizes the through-bond mode of transmission, while the term *field* emphasizes the through-space mode of transmission). Thus the greater acidity of trifluoroacetic acid than acetic acid is attributed to the inductive/field effect of the electron-withdrawing fluorine substituents that stabilize the trifluoroacetate anion.

The second mechanism, termed the *resonance* effect, is due to interactions between the substituent and the reaction center involving π-type systems on the substituent and substrate. Of course this mechanism can only operate when both the substituent and the substrate possess orbitals of π (or pseudo π) symmetry. For example, the stabilizing interaction of an odd electron on a carbon atom with an adjacent double bond, as in the allyl reaction, exemplifies the resonance effect (discussed in Section 2.2.6).

In order for the σ scale to be able to correlate rate and equilibrium data, the relative contribution of the two mechanisms of transmission (inductive/field and resonance) must be the same in the reaction under consideration and in the model reaction (benzoic acid ionization). However, a moment's thought shows that in some cases there will be deviations from this pattern. In the benzoic acid ionization reaction, the substituent effect is primarily due to interaction of the substituent X with the negative charge in the substituted benzoate ion (**6.1**). In this species no direct resonance interaction between the negative charge and the substituent is possible, though the substituent does transmit some of its effect through the π system of the benzene ring. This general situation, where there is no direct resonance interaction between the substituent and the reaction center, is very common, and that is why the σ scale is widely applicable. However, it also follows that the σ scale should only be applicable for that general group of reactions.

6.1

If in some reaction the substituent *can* come into direct resonance interaction with the reaction center, then the balance of inductive and resonance contributions will change significantly from that in the benzoic acid system, and *resonance interactions will become more important*. In such circumstances correlations with the σ scale are unlikely to occur.

Consider, for example, the *p*-methoxy-*tert*-cumyl cation.

6.2   **6.3**

In this species the cationic center is adjacent to the aromatic ring, so substituents, such as *p*-CH_3O, can undergo *direct* resonance interaction with the cationic site, as shown in the resonance hybrid **6.3**. Therefore such substituents will be particularly effective in stabilizing the positive center, much more so than reflected in their σ_{para} value. So for reactions involving substituted cumyl cations, the σ scale would be unsuitable. *An alternative model reaction, in which positive charge is generated either in, or adjacent to, the aromatic ring, needs to be found* and used to measure the substituent effects. In fact the reaction that was chosen is actually based on the family of *tert*-cumyl derivatives.

6.2.2 σ^+ Scale

The reaction chosen to model reactions in which positive charge is generated either in, or adjacent to, the aromatic ring is the solvolysis of substituted *tert*-cumyl chlorides in 90% acetone–water.

$$ArC(CH_3)_2Cl \xrightarrow{\text{90% acetone–H}_2\text{O}} \left[ArC(CH_3)_2^+\right] \xrightarrow{\text{H}_2\text{O}} ArC(CH_3)_2OH \quad (6.3)$$

Since the rate constant for this reaction depends on the stability of the inter-mediate cation formed, the rate constants for this reaction set may be used to generate a new set of substituent parameters called σ^+. Values of σ^+ for a range of substituents are listed in Table 6.1. Note that entries only appear for those substituents that can act as π donors (such as NH_2, OCH_3, and the halogens, which possess a lone-pair orbital, and alkyl groups that can donate charge into the π system through hyperconjugation), and only for substituents at the para position, where direct resonance interaction with the reaction site can take place. Correlation with σ^+ for substituents that are not π donors, or for substituents at the meta position, simply utilize the regular σ values since in these cases the direct resonance effects that cause the difference in the two scales do not occur.

6.2.3 σ^- Scale

Just as reactions where a positive charge is generated adjacent to the ring (or, in the ring) resulted in the need for a special scale, reactions where *negative* charge is generated adjacent to the ring also require a separate scale. For example, the basicity of the phenoxide ion is greatly reduced by π acceptors, such as NO_2, due to the following resonance interaction:

6.4

The direct resonance interaction of the nitro group with the negative charge on O^- means that the nitro group is a much more effective electron-withdrawing group than the σ scale would suggest. Therefore a special scale that takes this enhanced resonance interaction into account is required for reactions where direct stabilization of *negative* charge can occur. Such a scale is obtained from the pK_a values for the ionization of substituted phenols.

$$ArOH \rightarrow ArO^- + H^+ \quad (6.4)$$

The resulting scale is termed the σ^- scale, and values of σ^- are listed in Table 6.1. Note that entries are restricted to π acceptors, such as NO_2, CN, and $COCH_3$, since it is for these substituents that differences between σ_{para} and σ^- are most pronounced. Some examples of reactions whose rate (or equilibrium) constants correlate with σ^+ or σ^- are listed in Table 6.2.

6.2.4 Use of the Hammett Equation

The Hammett equation is an important tool in physical organic chemistry. Since its inception the Hammett equation has been applied on countless occasions to an enormous number of different reactions. Its wide application stems from the fact that Hammett correlations can provide important mechanistic information and predict unknown rate and equilibrium constants.

Kinetic ρ values, obtained from plots of *rate* constants against the appropriate σ parameters, provide useful information on reaction mechanism. The *magnitude* of the ρ value indicates whether there is substantial charge development in the transition state, and the *sign* of the ρ value indicates whether the developing charge is positive or negative. In addition, the particular σ scale that is used, say σ or σ^+, provides some indication as to whether the charge that is formed is adjacent to the aromatic ring or more remote.

Consider, for example, the solvolysis of benzhydryl chloride in ethanol (a reaction listed in Table 6.2). The large negative ρ value for that reaction (-5.1) reflects the fact that the reaction rate is greatly enhanced by *electron-releasing* substituents. This result in turn suggests that substantial *positive* charge develops on the benzylic carbon in the transition state of that reaction, as shown in **6.5**.

6.5

The fact that the rate constant correlates with σ^+, rather than σ, provides further evidence that the positive charge is formed adjacent to the ring (thereby enabling π-donor substituents to undergo direct resonance interaction with the developing charge). In other words, the Hammett correlation for this reaction points to an S_N1 mechanism in which C—Cl bond heterolysis is rate determining, rather than an alternative mechanism, such as direct S_N2 substitution of chloride ion by the solvent.

A sudden break in a Hammett plot, as shown in Fig. 6.2 for the hydrolysis of ethyl benzoate in 99.9% sulfuric acid, signals that something unusual is taking place in this reaction family. In this case the break signifies a *mechanistic*

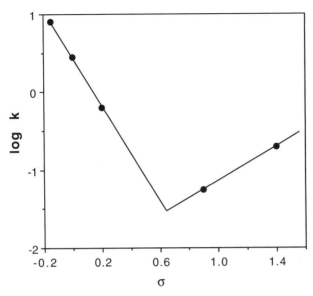

Fig. 6.2. Plot of log k for hydrolysis of ring-substituted ethyl benzoates in 99.9% sulfuric acid against the σ parameter. Adapted from J. E. Leffler and E. Grunwald, *Rates and Equilibria of Organic Reactions*, Wiley, New York, 1963.

changeover that is induced by the substituent variation. For electron-releasing substituents (low σ values) the Hammett plot has a *negative* slope. This suggests a mechanistic pathway in which *positive* charge develops in the transition state of the reaction (or, more precisely, a transition state where the charge change is in a positive direction). A mechanism that is consistent with this observation is the formation of a substituted benzoyl cation, as shown in Eq. (6.5), which then reacts with water to give the reaction product; in this mechanistic pathway the rate-determining heterolysis of the acyl–oxygen bond is facilitated by electron-releasing substituents, since a positively charged carbon atom is generated adjacent to the ring.

$$(6.5)$$

For electron-withdrawing groups (large σ values) the Hammett plot has a *positive* slope. This suggests the mechanistic pathway

$$(6.6)$$

The positive slope suggests either *negative* charge formation in the transition state, or alternatively, a reduction in reactant *positive* charge. In the present case it is the latter possibility. The electron-withdrawing groups, by *destabilizing* the positively charged reactant ion, enhance the rate of hydrolysis so a positive ρ value is observed for these substituents.

Why does the mechanism change for different substituent types? Both mechanistic pathways involve bond fission [either acyl–oxygen fission [Eq. (6.5)] or alkyl–oxygen fission [Eq. (6.6)]. If we assume in both cases that a carbocation is formed (either substituted benzoyl cation from the acyl–oxygen fission, or ethyl cation from the alkyl–oxygen fission), then the favored pathway in any given case will be the one in which the *more* stable cation is formed. The experimental results tell us that for the more electron-releasing substituents, the substituted benzoyl cation is the more stable ion, while for the more electron-withdrawing substituents ($\sigma > 0.7$), the ethyl cation is the more stable ion. *A change in the substituent will change the mechanism when the relative stability of the two competing cations inverts.* Of course, the possibility that an S_N2 type process occurs, which bypasses ethyl cation formation, also needs to be considered so that the above interpretation is somewhat simplistic.

The above example, however, clearly demonstrates the power of the Hammett correlation. Considerable mechanistic information is uncovered from a set of simple kinetic measurements. In this case insight into the mechanistic pathways for the hydrolysis of different family members was obtained, as well as some indirect information regarding the solution stabilities of different carbocations.

The Hammet equation is an example of a *linear free energy relationship*. What is the basis for this terminology? Recall that free energy changes are proportional to log K and log k (see Section 5.4).

$$\Delta G^\circ = -2.303RT \log K \qquad (6.7)$$

$$\Delta G^\ddagger = -2.303RT \log (kh/kT) \qquad (6.8)$$

If we write the Hammett equation (6.2) in the form

$$\log (K_X'/K_H') = \rho \log (K_X/K_H) \qquad (6.9)$$

where K_X' and K_H' are the equilibrium constants for a particular reaction with substituents X and H, respectively, and σ has been replaced by the benzoic acid ionization constants K_X and K_H [Eq. (6.1)], then by substituting equilibrium constants for free energies into Eq. (6.9), according to Eq. (6.7), we obtain

$$\Delta G_X^{o\prime} - \Delta G_H^{o\prime} = \rho(\Delta G_X^{o} - \Delta G_H^{o}) \qquad (6.10)$$

where $\Delta G_X^{o\prime}$ and $\Delta G_H^{o\prime}$ are the free energy changes for the particular reaction with substituents X and H, respectively, and ΔG_X^{o} and ΔG_H^{o} are the free energy changes for benzoic acid ionization with substituents X and H, respectively.

Now we can see the basis for the term—linear free energy relationship. The Hammett relationship indicates *that the effect of a substituent change on the free energy of any reaction is directly proportional to the effect of that substituent change on the free energy of benzoic acid ionization*; the relationship is a linear one. The term applies not just to the thermodynamic form of the Hammett equation [Eq. (6.2a)] but to the kinetic form [Eq. (6.2b)] as well. An expression for substituent effects on *free energies of activation*

$$\Delta G_X^{\prime\ddagger} - \Delta G_H^{\prime\ddagger} = \rho(\Delta G_X^{o} - \Delta G_H^{o}) \qquad (6.11)$$

which is based on Eq. (6.8) (for rates) rather than on Eq. (6.7) (for equilibria), may also be derived. In fact *any* empirical relationship that quantitatively links the rate constants (or equilibrium constants) for one set of reactions and the rate constants (or equilibrium constants) for a second set of reactions, as described by one of the following:

$$\log k_1 = \alpha \log k_2 + C \qquad (6.12)$$

$$\log K_1 = \alpha \log K_2 + C \qquad (6.13)$$

$$\log k_1 = \alpha \log K_2 + C \qquad (6.14)$$

where α is a proportionality constant and C is a constant, will conform to Eq. (6.10) [or Eq. (6.11)] and belong to that general class of relationships termed linear free energy relationships (LFERs).

6.2.5 Polar Effects in Aliphatic Systems

The three scales of substituent parameters described till now (σ, σ^+, and σ^-), depend on the substituent relaying its effect to the reaction center through a benzene ring. As such, these scales cannot be used to assess substituent effects in aliphatic systems. Moreover, as discussed above, substituent effects transmitted through a benzene ring do so through a combination of polar (the simpler term for inductive/field) and resonance effects. Substituent effects in aliphatic systems, however, may operate via just a polar mechanism, since no π system— a prerequisite for the resonance mechanism to operate—may be present.

Several different approaches have been utilized in order to evaluate polar effects. The first of these was developed by Taft[1] and was based on the effect of substituents on the rate of hydrolysis of substituted acetate esters, XCH_2COOR. Taft argued that the transition states for the acid- and base-catalyzed reactions of a substituted acetate ester, presumably resemble the intermediates that are formed in the reaction—a cationic intermediate **6.6** and an anionic one **6.7**, respectively. Given that **6.6** is positively charged while **6.7** is negatively charged, the *difference* in the rates for these two reactions

$$
\begin{array}{cc}
\overset{\displaystyle OH}{\underset{\displaystyle \overset{+}{O}H_2}{X-\overset{|}{\underset{|}{C}}-OR}} & \overset{\displaystyle O^-}{\underset{\displaystyle OH}{X-\overset{|}{\underset{|}{C}}-OR}} \\[2em]
\textbf{6.6} & \textbf{6.7}
\end{array}
$$

was considered to reflect the polar effect of the substituent on the reaction center. Steric factors were presumed to be insignificant since the two structures differ by just two protons. Thus Taft defined a polar substituent parameter σ^* as

$$\sigma^* = \frac{1}{2.48}\left\{\log\left(\frac{k}{k_0}\right)_B - \log\left(\frac{k}{k_0}\right)_A\right\} \tag{6.15}$$

where k and k_0 are the substituted and unsubstituted rate constants for hydrolysis and the A and B subscripts denote acid- and base-catalyzed reactions, respectively. The factor of $1/2.48$ was introduced to make the σ^* scale size consistent with the σ scale. Note also that the reference substituent chosen was CH_3 rather than H (i.e., acetate esters were used as the reference substrates).

Some questions regarding the Taft procedure have been raised over time. For example, though the Taft parameter was devised to specifically eliminate steric effects, there has been some question as to how effective this has been. Also, in analogy with the σ scale, it seemed that H rather than CH_3 should be the appropriate reference substituent. Consequently several other approaches have been followed over the years, and an alternative scale, termed σ_I, has been developed. Values of σ_I have been obtained by measuring substituent effects on the acidity of 4-substituted bicyclo[2.2.2]octane-1-carboxylic acids (**6.8**), and 4-substituted quinuclidinium ions (**6.9**).

$$
\begin{array}{cc}
\textbf{6.8} & \textbf{6.9}
\end{array}
$$

These model systems seem particularly well designed since they involve a rigid bicyclic structure where the substituent-reaction center distance is fixed and steric effects are absent. Acidity differences for different substituents can indeed then be attributed to the electronic effect of the substituent on the reaction center. A weakness of these systems, however, is that a wide range of substituents is not available, so studies on the widely available α-substituted acetic acids have also been used. Values of σ_I for some common substituents are listed in table 6.1.

6.3 LIMITATIONS OF THE HAMMETT APPROACH

It has recently become apparent that even with the range of substituent parameters that are available, correlations between rates (or equilibria) and some substituent scales are not always observed. To resolve this difficulty *dual parameter substituent constants* were developed. This approach was pioneered by Yukawa and Tsuno[2] for benzene derivatives and then generalized to include other systems as well. The idea is that the σ constant for any substituent can actually be separated into its inductive/field and resonance components, each of which may be quantified. Thus,

$$\sigma = fF + rR \qquad (6.16)$$

where F and R are the inductive/field and resonance parameters, respectively, and f and r are factors that determine the weight of the two components in any given case. The F and R values may be found in substituent tables, while values of f and r need to be determined for each system, and this is done by a best-fit analysis using standard statistical procedures.

A concern with this approach is that additional (adjustable) parameters (f and r) need to be incorporated into the treatment. It is true that improvements in the correlation result when the dual substituent parameter procedure is used. However, it is important to verify that the improved correlation derives from an improved *physical* picture, and not simply from the incorporation of additional adjustable parameters. Ultimately, it must be remembered that the enormous impact of the Hammett equation derives from its generality; the introduction of additional substituent scales together with adjustable parameters whose purpose is to improve the quality of fit, to some extent undermines the usefulness of the Hammett approach.

A final point concerning the limitations of the Hammett approach: The idea that a substituent has a well-defined electronic effect that is largely independent of the substrate to which it is bound, must be treated as a simplifying assumption. In actual fact the electronic behavior of a substituent may be greatly modified by the substrate to which it is bound. For example, the CN substituent, which is considered to be both a σ and a π acceptor, *acts as a π donor* when bonded to a very powerful π-accepting substrate, such as a carbocation (in $NC-CH_2^+$). Simple orbital considerations, illustrated in Fig. 6.3, explain this

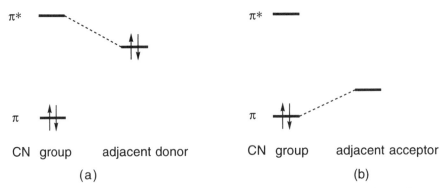

Fig. 6.3. Orbital interaction diagram showing the cyano group as both an electron acceptor and an electron donor. (a) Cyano as a π acceptor when adjacent to a donor group with a high-lying π type occupied orbital and (b) as a π donor when adjacent to a strong acceptor group with a low-lying π type unoccupied orbital, such as CH_2^+.

unexpected behavior. The CN group possesses a low-lying occupied π orbital, as well as a low-lying unoccupied π^* orbital. Normally the electronic nature of the group is governed by the π^* orbital so the group interacts with neighboring groups as a π *acceptor* (Fig. 6.3a). However, the presence of the occupied π orbital can, in principle, lead to the CN group acting as a π *donor*. This is likely to occur when the group is bonded to a substrate with a particularly low-lying vacant p- or π-type orbital, such as exists in C^+ (Fig. 6.3b). Therefore we must conclude that substituent effects generally *cannot* be assumed to be fixed and predetermined. Substituent effects in a particular substrate necessarily involves the mutual interaction of two groups. As such, the effect of one group on the other must, in the final analysis, depend to some degree on the nature of *both* groups; that is, both the magnitude *and* the direction of substituent effects are likely to be influenced by the substrate to which the substituent is bound.

Thus the wide application of the Hammett and Hammett-type equations and their enormous utility should be viewed as quite remarkable. The extraordinary generality of the Hammett relationship is empirical; simple theoretical considerations suggest that the approach cannot be general in a fundamental way. The implication would appear to be that the Hammett relationship applies within a limited spectrum of chemical behavior and that in extreme cases, when these bounds are exceeded, deviations from the normal pattern will increasingly manifest themselves.

6.4 INTRAMOLECULAR INTERACTIONS

To state that a substituent has a stabilizing effect on a particular substrate is common. For example, we say that a carbocationic center is strongly stabilized by an electron-releasing substituent such as CH_3. The effect of methyl substituents on carbocation stability is said to increase in the order

$$CH_3^+ < CH_3CH_2^+ < (CH_3)_2CH^+ < (CH_3)_3C^+$$

But what do we really mean by the statement that the *tert*-butyl cation is more stable than the methyl cation? When we compare the stabilities of two isomeric species, such as the 1-propyl and 2-propyl cation, the statement is clear. Since the two ions have the same molecular mass, their relative energies can be obtained through a comparison of their heats of formation, or the heats of some other standard reaction. But for two species of *different* molecular mass the concept of relative stability is more subtle and can only be assessed indirectly.

A way to quantify the relative stabilities of a set of related species (that are not isomeric) is through the use of *bond-exchange reactions*. A bond-exchange reaction may be written in general form as

$$X-Z + X'-Z' \rightarrow X'-Z + X-Z' \qquad (6.17)$$

and its energy provides a measure of intramolecular interactions. The energy of Eq. (6.17) will measure the difference in the interaction between X and Z on the one hand, and X and Z' on the other, using X' as a reference. An example makes this concept easier to understand.

Let us say we want to know how a substituent X affects a functional group Y within the molecule XCH_2Y. In that case we can measure the energy of the reaction

$$XCH_2Y + CH_4 \rightarrow CH_3X + CH_3Y \qquad (6.18)$$

On the left-hand side of this equation, X and Y are bonded to the same carbon, while on the right-hand side they are bonded to different carbon atoms. If the energy of this reaction is positive then there must be a *stabilizing* interaction between X and Y in XCH_2Y. Conversely, if the energy of this reaction is negative then there is a *destabilizing* interaction between X and Y. Of course, when we say that the interaction between X and Y is stabilizing or destabilizing, the statement is made with reference to the X--H and Y--H interactions in CH_3X and CH_3Y, since the reference compound chosen in this case is CH_4.

Let us return to our example of a CH_3 substituent effect on a carbocationic center. When we say that $CH_3CH_2^+$ is more stable than CH_3^+, what we really mean is that the energy of the reaction

$$CH_3-CH_2^+ + H-CH_3 \rightarrow H-CH_2^+ + CH_3-CH_3 \qquad (6.19)$$

is *positive*. The positive energy of this reaction tells us that *the interaction between a CH_3 group and the CH_2^+ moiety is more favorable than that between H and CH_2^+ (using CH_4 as a reference).* Thus Eq. (6.19) is a specific example of a general bond-exchange reaction.

Note that in order to characterize the effect of substituents on the energy of some species, it is always done with respect to some reference. Substituent effects on the energy of a species *cannot* be obtained in an absolute way; changing the reference necessarily changes the values of the stabilization energies. For this reason the reference compound is chosen in such a way that interactions between it and the substituents are minimized; CH_4 and H_2 are common reference compounds, though others may be appropriate in certain circumstances.

If we use such bond-exchange reactions the stabilization energies of the important organic intermediates (carbocation, radicals, carbanions, and carbenes) can be obtained by using ab initio MO calculations. The computational approach allows a wide body of data to be obtained by a single procedure, though modification of the data may occur as higher level calculations become available. Some results are listed in Table 6.3.

Substituent effects tend to be large for the charged intermediates (carbocations and carbanions). This occurs because the dispersal of charge is a primary

TABLE 6.3. Calculated Stabilization Energies for Carbocations, Carbon Radicals, Carbanions, and Carbenes (kcal mol^{-1})^a

X	XCH_2^{+b}	$XCH_2^{\cdot c}$	XCH_2^{-d}	XCH Singlet^e	XCH Triplet^f
H	0	0	0	0	0
Li	77.0	9.0	19.2	9.2	27.0
BeH	16.3	8.6	31.8	−1.6	18.5
BH_2	27.4	12.2	53.3	36.8	15.6
CH_3	35.5	2.5	−3.1	12.0	5.4
NH_2	99.6	11.4	−0.5	61.5	13.0
OH	66.3	8.9	5.6	51.2	11.0
F	25.9	4.7	10.5	35.0	5.5
SH	64.9	7.4	19.2	32^e	
Cl	31.0	4.5	17.4	16^e	
$CH{=}CH_2{}^g$	54	21	28	17	
$C{\equiv}CH^g$	33	15		14	
$C{\equiv}N^g$	−9	11	48	9	21
$CH{=}O^g$	−4	4	60	−1	17

^{a}Stabilization energies defined by the energy of

$$XY + CH_4 \rightarrow HY + CH_3X$$

where $Y = CH_2^+$, $CH_2^{\cdot}$, CH_2^-, CH (singlet), and CH (triplet). Data taken from P. v. R. Schleyer, *Pure Appl. Chem.* **59**, 1647 (1987) and from W. J. Hehre, L. Radom, P. v. R. Schleyer, and J. A. Pople, *Ab Initio Molecular Orbital Theory*, Wiley, New York, 1986, p. 348.
^{b}Data calculated at the MP4 SDTQ 6-31G*//6-31G* level.
^{c}Data calculated at the UMP2 6-31G*//6-31G* level.
^{d}Data calculated at the MP4/6-31+G*//6-31+G* level with ZPE.
^{e}First row: MP4/6-31G*//3-21G(*) level; Second row: 3-21G(*)//3-21G(*) level.
^{f}UMP4/6-31G*//3-21G level.
^{g}Data for $CH{=}CH_2$, $C{\equiv}CH$, $C{\equiv}N$ and $CH{=}O$ at the 3-21G//3-21G level.

mechanism for stabilization. We also learn from the data in Table 6.3 that substituent effects operate through both the σ and π frameworks. Thus carbocations are stabilized by both σ- and π-electron donors. The electropositive substituents, Li, BeH, and BH$_2$, are σ donors and stabilize the cationic center by donating charge through the σ framework, while substituents, such as NH$_2$, OH, F, and CH=CH$_2$, are π-type donors, and donate electron density that is located in a lone pair or π orbital, into the vacant C $2p$ orbital. For the substituents C$\equiv$N and CH=O, the *destabilizing* effect of σ withdrawal appears to dominate the stabilizing effect of π donation from an occupied π orbital.

Carbanions are stabilized by both σ-electron-withdrawing substituents, such as OH, F, Cl, and π-electron-withdrawing substituents. The π-electron-withdrawing substituents may be classified into two types: substituents, such as Li, BeH, and BH$_2$, which possess a vacant $2p$ orbital, and substituents, such as CN and CHO, which possess a low-lying vacant π^* orbital. Both types of substituent stabilize the carbanion by delocalizing the negative charge into a vacant orbital of π symmetry. The ΔG° values for the dissociation of carbon acids in the gas phase and in solution (listed in Table 8.2) are also a measure of substituent effects on carbanion stability, and show a similar pattern.

For the radical species XCH$_2^{\displaystyle \cdot}$ the absolute magnitudes of the stabilization energies are in most cases significantly smaller than for the charged species XCH$_2^+$ and XCH$_2^-$. For radicals, the important stabilization mechanism is the interaction of the radical center with a π system (vacant or occupied), such as those in BH$_2$, NH$_2$, and CH=CH$_2$. Interaction of the radical center with the π system leads to a more extended and delocalized π system, which may be readily depicted using either MO diagrams or VB resonance structures.

The stabilization energies of the singlet and triplet carbenes can be understood in the same terms. The singlet carbene, though it is uncharged, can be thought of as a species with both carbocationic *and* carbanionic character. It possesses a carbocationic-like vacant $2p$ orbital on carbon, and an orthogonal carbanionic-like lone-pair orbital on the same carbon. Both of these orbitals can interact strongly with substituent orbitals. Hence, substituent effects on singlet carbenes are surprisingly large for an uncharged species.

Triplet carbenes can be thought of as biradical species, and therefore stabilization energies are typically larger than for ordinary carbon radicals. There are *two* unpaired electrons that can interact with the substituent, as compared with the single unpaired electron in XCH$_2^{\displaystyle \cdot}$.

6.5 RATE–EQUILIBRIUM RELATIONSHIPS AND THE SIGNIFICANCE OF THE BRØNSTED PARAMETER

Factors that govern the rates of chemical reactions have always been more difficult to understand and predict than those that govern their equilibria. Thus there has been a long-standing interest in the empirical relationship between reaction rates and equilibria. The implicit hope has been that in some way the greater understanding associated with reaction equilibria may somehow be extended to reaction rates as well.

The earliest example of a rate–equilibrium relationship is the Brønsted catalysis law. In the 1920s Brønsted discovered that the rate constants of certain reactions that are catalyzed by acid are also dependent on the acid strength. The relationship between the two was found to be

$$\log k = \alpha \log K_a + C \tag{6.20}$$

where k is the rate constant for the reaction in question and K_a is the acidity constant for the catalyzing acid. An example of a reaction that obeys the Brønsted catalysis law is the acid catalyzed hydrolysis of vinyl ethers:[3]

$$ROCH{=}CH_2 + H_2O \xrightarrow{\;H^+\;} ROH + CH_3CHO \tag{6.21}$$

A plot of the log of the measured rate constant against the pK_a of the carboxylic acids used to catalyze the reaction gave a linear correlation whose slope, termed the *Brønsted slope* or *Brønsted parameter* and which provides a measure of α, was found to be 0.70. The significance of the slope of the correlation has been a subject of intense interest and controversy and will be discussed later.[4]

The Brønsted catalysis law is very general in its form, and as noted earlier, represents an example of a rate–equilibrium relationship [Eq. (6.14)] in which a set of rate constants k correlates with a set of equilibrium constants K. In some cases the rate and equilibrium constants refer to the *same* reaction; in others they refer to two *different* reactions (though the two reactions would need to be related). Let us illustrate this with an example.

When the rate and equilibrium constants for nucleophilic attack of substituted pyridines on methyl iodide

$$\tag{6.22}$$

were measured, it was found that a plot of the rate constants for different substituents X in the pyridine ring, against the corresponding equilibrium constants, gave a linear correlation whose slope α was about 0.26.[5]

Alternatively, it was possible to obtain a Brønsted plot by plotting the set of rate constants against the set of pyridinium ion pK_a values, since the protonation reaction

$$\tag{6.23}$$

is similar to the reaction in question, and the pK_a data are readily available. It is true that the pyridine nitrogen in Eq. (6.23) is protonated rather than methylated, as in Eq. (6.22). However, both reactions generate a full positive charge on nitrogen, and it is therefore assumed that the pK_a data may serve as a useful model for the equilibrium process. Indeed, plots of the rate constants for Eq. (6.22) against pyridinium ion pK_a values [Eq. (6.23)], shown in Fig. 6.4 for three different leaving groups, lead to Brønsted slopes of about 0.3—very close to the value of 0.26 obtained in the direct rate–equilibrium plot.

What do we learn from these Brønsted correlations? The correlations tell us that rates and equilibria for these reactions are affected by substituents in the same way, *but that rates are less sensitive to substituent changes than equilibria* (~30% as sensitive). This behavior is typical. Many reactions exhibit linear rate–equilibrium relationships in which the rates are less sensitive to substituent changes than the equilibria. Another way of saying this is that the magnitude of the Brønsted parameter is commonly (but not always) found to lie in the range 0–1. What is the *mechanistic* significance of this parameter?

The significance of the Brønsted parameter has been a subject of considerable controversy. In 1953 Leffler[6] made a far-reaching mechanistic interpretation of the Brønsted parameter. He proposed that *the magnitude of the parameter provided an experimental measure of the location of the transition state along the reaction coordinate.* In other words a Brønsted parameter of about 0.3 for the reaction shown in Eq. (6.22) suggests that the transition state is relatively early and has progressed just 30% of the way along the reaction coordinate. Specifically, this means that at the transition state the C—N bond is 30% formed and that the developing charge on the pyridine N atom is about +0.3, as illustrated in

$$0.3\ \delta+ \qquad 0.3\ \delta-$$
$$\text{N--------R---X}$$

6.10

Leffler's proposal was based on the assumption that perturbations (induced by substituent or solvent changes) on the free energy of the transition state will be related to those on reactants and products according to

$$\delta G^{\ddagger} = \alpha \delta G_{\text{P}} + (1 - \alpha)\delta G_{\text{R}} \qquad (6.24)$$

Here δ is an operator that indicates the change introduced into the function that it precedes as a result of some perturbation; $G^{\ddagger}$, G_{P}, and G_{R} are the free energies of the transition state, products, and reactants, respectively; and α is a mixing factor whose value lies in the range 0–1. Thus Leffler proposed that perturbations on the free energies of reactants and products will only be *partially* manifested in the transition state and the extent of the perturbation on the transition state will depend on the degree to which the transition state has

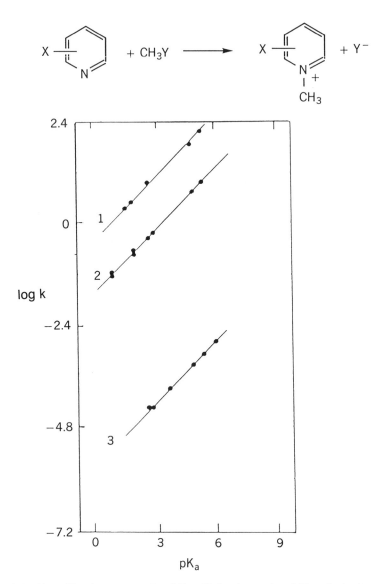

Fig. 6.4. Plot of log k versus nucleophile pK_a for the nucleophilic substitution reaction of CH_3Y by a set of 3- and 4-substituted pyridines at 25°C. Y = OCCl (line 1), Y = SO_3F (line 2), and Y = I (line 3). [Reprinted in part with permission from E. M. Arnett and R. Reich, *J. Am. Chem. Soc.* **102**, 5892 (1980), Fig. 9. Copyright © (1980) American Chemical Society.

reactant-like and product-like character. This, in turn, is governed by the magnitude of α so that α may be viewed as *a measure of transition state structure.* Values of α close to 0 indicate a reactant-like transition state, while values close to 1 would indicate a product-like transition state.

By substituting $\Delta G^{\ddagger}$ for $G^{\ddagger} - G_R$ and $\Delta G°$ for $G_P - G_R$, Eq. (6.23) may be rearranged into a more convenient form

$$\delta\Delta G^{\ddagger} = \alpha\delta\Delta G° \qquad (6.25)$$

This formulation makes it clear that changes in the free energy of reaction will only be *partially* reflected in the free energy of activation. Since free energy changes are directly related to rate and equilibrium constants [by Eqs. (6.7) and (6.8)], the free energy relationship of Eq. (6.25) may be converted into the generalized rate–equilibrium relationship, Eq. (6.14).

An energetic representation of the Leffler assumption is shown in Fig. 6.5. The effect of a perturbation on the free energy of the product $\delta\Delta G°$ is illustrated by the broken line. It can be seen that this perturbation has no effect on the free energy of reactants, but as one proceeds along the reaction coordinate the reaction complex is increasingly stabilized, compared to the unperturbed system. Thus the extent to which the transition state will be stabilized will depend on the position of the transition state along the reaction coordinate.

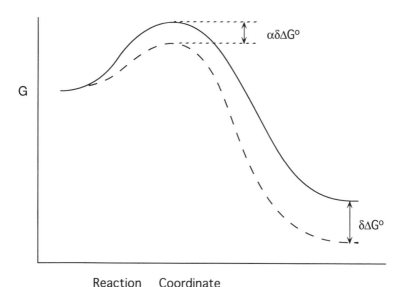

Fig. 6.5. Schematic illustration of the Leffler assumption [Eq. (6.24)], which shows how perturbations on the free energy of a reaction $\delta\Delta G°$ are only partially reflected in the free energy of the transition state $\alpha\delta\Delta G°$, where α lies in the range 0–1.

The Leffler assumption is seductively appealing. However, with time it has become increasingly apparent that there is some difficulty with this description. The first serious challenge to the approach came about with the observation by Bordwell and co-workers[7] that the deprotonation reaction of substituted nitroalkanes

$$
HO^- + H-\underset{\underset{CH_3}{|}}{\overset{\overset{R}{|}}{C}}-NO_2 \longrightarrow \left[\overset{\delta-}{HO} \text{---} H \text{-----} \underset{\underset{CH_3}{|}}{\overset{\overset{R}{|}}{C}} \overset{\delta-}{-NO_2} \right]
$$

$$
HO-H + \underset{H_3C}{\overset{R}{\diagdown}}C=NO_2^- \qquad (6.26)
$$

while still conforming to a rate–equilibrium relationship, resulted in Brønsted slopes that were *greater than 1* (typically ~ 1.5). In other words, in this particular case, rates are *more* sensitive to substituent effects (changes in the R group) than equilibria. So the interpretation of α as a measure of the position of the transition state along the reaction coordinate clearly has limitations. A molecular interpretation of this result would be that there is *more* negative charge localized on C in the transition state than in the product, even though this is in conflict with the Leffler assumption that the transition state is intermediate in character between reactants and products. A buildup of negative charge on C in the transition state would explain the value of $\alpha > 1$.

An even more extreme case of an anomalous rate–equilibrium relationship is provided by the family of identity reactions

$$
^*Cl^- + CH_2-Cl \qquad \qquad ^*Cl-CH_2 + Cl^-
$$

$$
\longrightarrow \qquad (6.27)
$$

In this case the equilibrium constant must be 1 regardless of the substituent X in the aromatic ring, since the reaction is an identity reaction, while the rate constants for exchange are variable and dependent on the nature of the X. *Consequently, the rate–equilibrium plot for this reaction must be a straight line of infinite slope!* Clearly, this particular anomalous α value provides *no* information on the structure of the transition state of this reaction, and warns us that the molecular interpretation of Brønsted parameters needs to be made cautiously.

So how valid *is* the Leffler assumption? If we view the transition state in terms of a mixture of contributing configurations (as described in Section 4.4) it becomes clear that the Leffler assumption is not likely to be widely applicable. Typically three configurations, Ψ_R, Ψ_P, and Ψ_I (representing reactant, product, and intermediate configurations), are likely to be involved in a basic description of an energy profile, and the transition state of a reaction. Mixing in of Ψ_I into the transition state description necessarily means that the transition state will possess electronic character that is either *absent* (or, at least greatly reduced) in both reactant and product. For example, in an S_N2 reaction, the carbon that undergoes nucleophilic attack may experience either positive or negative charge buildup in the transition state, depending on the nature of the substituent on that carbon; electron-withdrawing groups on the central carbon will induce *negative* charge buildup, while electron-releasing groups will induce positive charge buildup (see Chapter 9). Also, it is well known that radical reactions often display polar character (due to the mixing in of a charge-transfer configuration; see Chapter 10) that is absent in both reactants or products.

The above considerations indicate that the Leffler assumption is not generally applicable. This then raises questions regarding the methodology that uses the Brønsted parameter as a measure of transition state structure. Note also that the term transition state structure is itself ambiguous. The transition state has both a *geometric* and an *electronic* structure and the two do not necessarily progress along the reaction coordinate in a concerted fashion; a transition state that is 20% advanced in geometric terms may be 50% advanced in electronic terms. So *does* the Brønsted parameter provide *any* mechanistic information?

It turns out that though the Brønsted parameter should not be used as an *absolute* measure of transition state structure, it may in some cases be used as a *relative* measure of transition state charge development. Since the transition state of a chemical reaction may be described by the mixing (or resonance hybrid) of a number of electronic configurations, the Brønsted parameter may provide information regarding the relative importance of the different contributors for different substitution patterns. This type of application for the nucleophilic substitution reaction of substituted benzyl derivatives is discussed in detail in Chapter 9.

Two general conclusions may be drawn.

1. Anomalous Brønsted parameters are *not* unusual or atypical; rates may be more sensitive to substituent effects than equilibria (i.e., $\alpha > 1$) whenever the transition state displays electronic character that is absent in reactants and products (due to some intermediate configuration Ψ_I). Values of $\alpha > 1$ are likely to be observed when the substituent is located in a site in the system that affects the energy of this intermediate configuration.

2. The interpretation of *normal* Brønsted parameters (i.e., $\alpha < 1$) in terms of the geometric location of the transition state along the reaction coordinate (described in Section 5.8.2) needs to be questioned, since the

Leffler assumption on which the idea is based is often inapplicable. However, it does seem that a comparison of Brønsted parameters for related systems may provide a *relative* measure of transition state charge development in certain cases (discussed in Chapter 9).

REFERENCES

1. R. W. Taft, Jr. In *Steric Effects in Organic Chemistry*, M. S. Newman, Ed., Wiley, New York, 1956, Chapter 13.
2. Y. Yukawa and Y. Tsuno, *Bull. Chem. Soc. Jpn.*, **32,** 971 (1959).
3. A. J. Kresge, H. L. Chen, Y. Chiang, E. Murrill, M. A. Payne, and D. S. Sagatys, *J. Am. Chem. Soc.* **93,** 413 (1971).
4. For a recent discussion on the significance of the Brønsted parameter, see A. Pross and S. S. Shaik, *New J. Chem.* **13,** 427 (1989).
5. E. M. Arnett and R. Reich, *J. Am. Chem. Soc.* **102,** 5892 (1980).
6. J. E. Leffler, *Science*, **117,** 340 (1953); J. E. Leffler and E. Grunwald, *Rates and Equilibria of Organic Reactions*, Wiley, New York, 1963.
7. F. G. Bordwell, W. J. Boyle, Jr., and K. C. Yee, *J. Am. Chem. Soc.* **92,** 5926 (1970); F. G. Bordwell and W. J. Boyle, Jr., *J. Am. Chem. Soc.* **94,** 3907 (1972).

7

IONIZATION POTENTIALS, ELECTRON AFFINITIES, AND SINGLET–TRIPLET EXCITATIONS

One of the simplest processes that an atom or molecule can undergo is to give up, or to accept, an electron. For this reason the energy associated with each of these two processes, termed the ionization potential I [Eq. (7.1)], and the electron affinity A [Eq. (7.2)], of a species are fundamental chemical properties.

$$M \rightarrow M^+ + e^- \qquad \Delta E = I \qquad (7.1)$$

$$M + e^- \rightarrow M^- \qquad \Delta E = -A \qquad (7.2)$$

While at first sight it may appear that these two processes are not of special chemical interest, this is far from the case. As discussed in Chapter 4, all chemical reactions involve a reorganization of the electrons about the atoms that comprise the reacting molecules. So *adding* an electron to a molecule, or *removing* an electron from a molecule, are fundamental processes that are intimately related to the reorganization that occurs in a chemical reaction. Indeed, as discussed in Chapter 9, many of the basic reactions of organic chemistry—nucleophilic addition and substitution, electrophilic addition and substitution, and so on, are under a camouflaged exterior, essentially single-electron shift processes. It directly follows that ionization potentials and electron affinities are basic parameters that are closely linked to chemical reactivity.

7.1 IONIZATION POTENTIALS

Ionization of a molecule to generate the corresponding cation and a free electron can be induced by either photoionization or electron impact. In the former a high-energy photon is responsible for ejecting the electron from the molecule, while in the latter it is a beam of high-energy electrons that provides the energy for ionization. Because quite often a molecule and its corresponding cation have different geometries, an important principle, termed the Franck–Condon principle, governs the actual ionization process. The Franck–Condon principle states *that during any electronic reorganization, the nuclear geometry can be considered to be frozen, because electronic motion is much more rapid than nuclear motion.* What this principle suggests therefore is that the *most probable* ionization transition will leave the resulting cation with the same nuclear geometry as the neutral parent. An energy diagram that illustrates the ionization process is shown in Fig. 7.1a.

Energy curves as a function of geometric distortion are shown for the neutral molecule (lower curve) and the cation (upper curve). It can be seen that the minima for the molecule and the ion are located at *different* points on the geometric coordinate. This difference simply indicates that in most cases the parent molecule and the cation have different equilibrium geometries. How then does the ionization process occur?

The process of ionization involves moving from the low-energy neutral curve up to the higher energy cation curve. The Franck–Condon principle tells us that the most probable ionizing transition is a *vertical transition* (labeled I_V) in which *no* geometric change takes place. However, as seen in Fig. 7.1a, such a transition leads to the formation of an ion whose geometry is displaced from its equilibrium position—or more precisely, the vertical transition leads to the

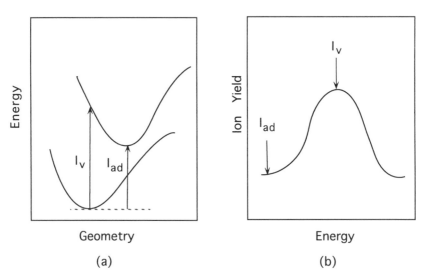

Geometry   Energy

(a)   (b)

Fig. 7.1. (a) Energy diagram showing adiabatic (I_{ad}) and vertical (I_V) ionization potentials for a molecule in terms of energy curves for the neutral molecule and the cation. (b) Unresolved peak in a photoelectron spectrum showing the adiabatic ionization potential (at peak onset), and the vertical ionization potential (at the peak maximum).

formation of a *vibrationally excited ion*. The energy of this transition is termed the *vertical ionization potential* I_V. The lowest energy that is actually required to ionize the molecule is termed the *adiabatic ionization potential* I_{ad}, and is defined by the difference in enthalpy between the parent molecule and its cation, *each at their equilibrium geometries*. The magnitude of I_{ad} is also depicted in Fig. 7.1a. It is clear from the diagram that the adiabatic ionization potential is *smaller* than the vertical ionization potential. The difference will depend on the difference in the geometries of the parent molecule and the cation, as well as the steepness of the cation distortion curve.

Both adiabatic and vertical ionization potentials may be measured experimentally. Mass spectrometric studies on equilibria such as

$$A + B^+ \rightleftharpoons A^+ + B \qquad (7.3)$$

provide adiabatic ionization potentials, since all ions are at equilibrium, that is, in their relaxed geometry. On the other hand, photoelectron spectroscopy may be used to obtain both vertical and adiabatic values. A typical unresolved peak in a photoelectron spectrum has the form shown in Fig. 7.1b. Peak onset (where the transition probability is low) provides a measure of the adiabatic ionization potential, while the peak maximum (where the transition probability is highest) provides a measure of the vertical ionization. Some vertical and adiabatic ionization potentials for selected molecules are listed in Table 7.1. Adiabatic ionization potentials for selected hydrocarbons and radicals are listed in Tables 7.2 and 7.3, respectively.

TABLE 7.1. Vertical (I_V) and Adiabatic (I_{ad}) Ionization Potentials of Selected Molecules[a]

Molecule	I_V (kcal mol^{-1})	I_{ad} (kcal mol^{-1})	Molecule	I_V (kcal mol^{-1})
NH_3	250.2 (10.8)	234.3 (10.2)	H_2O	291.0 (12.6)
CH_3NH_2	222.8 (9.7)	205 (8.9)	CH_3OH	252.7 (11.0)
$CH_3CH_2NH_2$	217.4 (9.4)	202 (8.8)	CH_3CH_2OH	244.9 (10.6)
$CH_3(CH_2)_2NH_2$	215.6 (9.3)	197 (8.5)	CH_3OCH_3	232.9 (10.1)
$CF_3CH_2NH_2$	241.2 (10.5)	224 (9.7)	$(CH_3CH_2)_2O$	222.8 (9.7)
$(CH_3)_2NH$	205.9 (8.9)	188 (8.2)	CH_2O	250.9 (10.9)
$(CH_3)_3N$	196.7 (8.5)	179 (7.8)	CH_3CHO	235.4 (10.2)
Quinuclidine	185.6 (8.0)	173 (7.5)	CH_3COCH_3	223.9 (9.7)
Pyridine	221.4 (9.6)		$HCOOCH_3$	254.1 (11.0)
Piperidine	199.7 (8.7)	181 (7.8)	CH_3COOH	250.2 (10.8)
$(CH_3)_2NCHO$	217.5 (9.4)		$HCOOH$	256.4 (11.1)
PH_3	244.4 (10.6)	229.6 (10.0)	H_2S	241.7 (10.5)
H_2Se	227.8 (9.9)		CH_3SH	217.7 (9.4)

[a]Data from D. H. Aue and M. T. Bowers. In *Gas Phase Ion Chemistry*, Vol. 2, Chapter 9, M. T. Bowers, Ed., Academic, New York, 1979. Values in electronvolts are shown in parentheses.

TABLE 7.2. Adiabatic Ionization Potentials of Selected Hydrocarbons[a]

Hydrocarbon	I_{ad} (kcal mol^{-1})	Hydrocarbon	I_{ad} (kcal mol^{-1})
Methane	288 (12.5)	Propene	224 (9.7)
Ethane	266 (11.5)	(E)-2-Butene	210 (9.1)
Propane	253 (11.0)	(Z)-2-Butene	210 (9.1)
Neopentane	235 (10.2)	1-Butene	221 (9.6)
Cyclohexane	227 (9.8)	Benzene	213 (9.2)
Ethylene	242 (10.5)	Toluene	203 (8.8)
Acetylene	263 (11.4)	P-Xylene	195 (8.5)
2-Butyne	221 (9.6)	Naphthalene	188 (8.2)

[a]Data from S. G. Lias, J. E. Bartmess, J. F. Liebman, J. L. Holmes, R. D. Levin, and W. G. Mallard, *J. Phys. Chem. Ref. Data.* **17**, Suppl. 1 (1988). Values in electronvolts are shown in parentheses.

7.2 ELECTRON AFFINITIES

The experimental determination of electron affinities is more difficult than that of ionization potentials, so most of the developments in this area have taken place in the last 15–20 years. As with ionization potentials, electron affinities may refer to either vertical or adiabatic values, though most reported data refer to adiabatic values. While ionization potentials for neutral molecules are always positive (it always requires energy to remove an electron from a molecule), electron affinity values may be either positive or negative. A negative electron affinity implies that the resulting anion is unstable with respect to electron detachment and has a very short lifetime (microseconds or less). Electron affinities for selected molecules are listed in Table 7.4. Note that the magnitudes

TABLE 7.3. Adiabatic Ionization Potentials of Selected Carbon Radicals[a]

Radical	I_{ad} (kcal mol^{-1})	Radical	I_{ad} (kcal mol^{-1})
Methyl	226.9 (9.8)	Fluoromethyl	205.2 (8.9)
Ethyl	193.3 (8.4)	Difluoromethyl	201.5 (8.7)
1-Propyl	186.8 (8.1)	Trifluoromethyl	211.5 (9.2)
2-Propyl	174.1 (7.5)	Vinyl	206.4 (9.0)
tert-Butyl	154.5 (6.7)	Allyl	186 (8.1)
Cyclopentyl	172.3 (7.4)	Benzyl	167.6 (7.3)
1-Butyl	184.7 (8.0)		
2-Butyl	170.9 (7.4)		

[a]Data from D. H. Aue and M. T. Bowers. In *Gas Phase Ion Chemistry*, Vol. 2, Chapter 9, M. T. Bowers, Ed., Academic, New York, 1979. Values in electronvolts are shown in parentheses.

TABLE 7.4. Electron Affinities of Selected Molecules[a]

Molecule	A (kcal mol^{-1})	Molecule	A (kcal mol^{-1})
Benzene	−26.3 (−1.14)	Nitrobenzene	>16.1 (>0.7)
Naphthalene	−4.6 (−0.20)	CF$_3$Br	21.0 (0.91)
Pyridine	−14.3 (−0.62)	CF$_3$I	32.3 (1.4)
Maleic anhydride	32.3 (1.4)		36.2 (1.57)
p-Benzoquinone	41.5 (1.8)	CH$_3$I	4.6 (0.2)
Tetracyanoethylene	53.0 (2.3)	CH$_3$NO$_2$	10.1 (0.44)
TCNQ[b]	64.6 (2.8)	Br$_2$	57.7–60.4 (2.51–2.62)
		Cl$_2$	53–56.3 (2.32–2.45)
		I$_2$	55.7–59.5 (2.42–2.58)

[a]Data from B. K. Janousek and J. I. Brauman. In *Gas Phase Ion Chemistry*, Vol. 2, Chapter 10, M. T. Bowers, Ed., Academic, New York, 1979. Values in electronvolts are shown in parentheses.
[b]TCNQ is tetracyanoquinonedimethane.

of electron affinities are substantially smaller than those for ionization potentials. Ionization potentials are typically about 200 kcal mol^{-1} (~ 9 eV) while the electron affinities of many molecules are close to 0 (normally <10 kcal mol^{-1}). Only molecules with strong electron-withdrawing groups have larger electron affinities and these can then range up to 60 kcal mol^{-1}.

Radicals derived from stable anions have significant electron affinities. Some examples are listed in Table 7.5.

TABLE 7.5. Electron Affinity of Selected Radicals[a]

Group	A (kcal mol^{-1})	Group	A (kcal mol^{-1})
F	78 (3.4)	CN	89 (3.9)
Cl	83.4 (3.6)	CCH	67.8 (2.9)
Br	78 (3.4)	NH$_2$	19.4 (0.8)
I	70.6 (3.1)	PH$_2$	29 (1.3)
OCOCH$_3$	78.2 (3.4)	H	17.4 (0.8)
OH	44 (1.9)	CH$_2$Ph	20.4 (0.9)
OCH$_3$	36.7 (1.6)	CH$_2$CN	35 (1.5)
SH	53.5 (2.3)	CH$_3$	1.8 (0.08)
SCH$_3$	43.4 (1.9)	OOH	27.5 (1.2)
SPh	56.9 (2.5)	CH$_2$COCH$_3$	41.2 (1.8)
SeH	51 (2.2)		

[a]Data taken from compilation in S. S. Shaik, H. B. Schlegel, and S. Wolfe, *Theoretical Aspects of Physical Organic Chemistry. The S$_N$2 Mechanism*, Wiley-Interscience, New York, 1992, Table A4.1, p. 155. Values in electronvolts are shown in parentheses. These data may also be used as ionization potential data for the corresponding anions, X$^-$.

7.3 CHARGE-TRANSFER COMPLEXES

One of the more common chemical applications of configuration mixing is found in the theoretical description of charge-transfer complexes.[1,2] It has long been known that good donor molecules, that is, those possessing low ionization potentials (typically <9 eV), such as alkylbenzenes, amines, and polycyclic aromatic hydrocarbons, when mixed together with good acceptor molecules, that is, those possessing high electron affinities, (typically 1–3 eV), such as tetracyanoethylene, iodine, and polynitrobenzenes, form 1 : 1 complexes, called charge-transfer complexes. The key feature of these complexes is that there is some bonding interaction between the donor and the acceptor, leading to the formation of a structurally stable species. These complexes are characterized by a low-frequency absorption band in the UV or visible range that is not observed in either of the isolated donor or acceptor molecules. The bonding interaction in these complexes, as well as their spectroscopic characteristics, were first explained by Mulliken.[1]

The bonding between donor and acceptor to generate a charge-transfer complex is different from the "normal" two electron bond on which much of organic structure is based. The nature of the bond may be understood in terms of *configuration mixing*. According to Mulliken, for a donor and acceptor pair (represented by D and A, respectively), two important electronic configurations need to be considered. These are DA (termed the no-bond configuration) and D^+A^- (termed the charge-transfer configuration). Formation of the charge-transfer complex can be understood by considering the configuration mixing diagram illustrated in Fig. 7.2, of the kind described in Chapter 3.

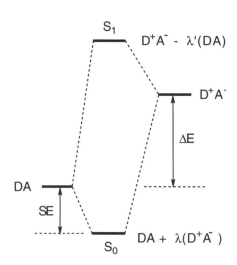

Fig. 7.2. Configuration-mixing diagram showing the formation of ground (S_0) and excited (S_1) states of a charge-transfer complex from the mixing of DA and D^+A^- configurations. SE is the stabilization energy of the complex.

Interaction of the two configuration wave functions DA and D^+A^- leads to the generation of two improved state functions—one for the ground state S_0, described by

$$S_0 = DA + \lambda(D^+A^-) \tag{7.4}$$

and one for the excited state S_1, described by

$$S_1 = D^+A^- - \lambda'(DA) \tag{7.5}$$

According to the equations governing the interaction of configurations (Section 3.4), the stabilization energy of the complex SE is given by

$$SE = B^2/\Delta E \tag{7.6}$$

where B, the interaction parameter [defined by Eq. (3.9)], is proportional to the overlap between the donor HOMO and the acceptor LUMO, and ΔE is the energy gap between the interacting configurations, which is equal to $I_D - A_A$. (In orbital terms this energy gap is just the HOMO–LUMO gap.)

Indeed, the Mulliken model for charge-transfer complexes explains several key characteristics of these complexes including (1) their stability, (2) the frequency of the charge-transfer band, (3) their structure, and (4) the electronic character of the ground and excited states of the complex.

1. It has been observed experimentally that good donor–acceptor pairs form more stable complexes than poor donor–acceptor pairs. For example, in Table 7.6, we see that the enthalpy of formation $\Delta H°$ for a series of amine–iodine complexes ranges from -4.8 kcal mol^{-1} for ammonia, the poorest donor, to -12.1 kcal mol^{-1} for trimethylamine, the most powerful donor. Similarly, an increase in the acceptor electron affinity also increases the stability of the complex. Thus for a series of polysubstituted nitrobenzenes, complex stability

TABLE 7.6. Relationship between the Heats of Formation of a Series of Amine–Iodine Charge-Transfer Complexes in *n*-Heptane at 20°C and the Amine Vertical Ionization Potential

Donor	I_V (kcal mol^{-1})	$\Delta H°$ (kcal mol^{-1})[a]
Ammonia	250.2	−4.8
Methylamine	222.8	−7.1
Ethylamine	217.4	−7.2
Dimethylamine	205.9	−9.9
Trimethylamine	196.7	−12.1

[a]Data for $\Delta H°$ taken from R. Foster, *Organic Charge-Transfer Complexes*, Academic, 1969, p. 191.

for a given donor increases in the order nitrobenzene < 1,4-dinitrobenzene < 1,3,5-trinitrobenzene < 1,2,3,5-tetranitrobenzene: the order of increasing electron affinity.

This dependence of stability on ionization potential and electron affinity is predicted by the Mulliken model. According to Eq. (7.6), the stability of the complex is inversely proportional to the energy gap ΔE. Given that ΔE is equal to $I_D - A_A$, we would expect that good donors (low I) and good acceptors (high A) will lead to the formation of more stable complexes, as is observed.

2. According to Fig. 7.2, photoexcitation of the ground-state complex S_0 will induce a transition to the excited state S_1. Indeed, charge-transfer complexes in their ground state normally show a strong absorption band in the UV or visible range. Since the ground state of the complex is described mainly by the DA configuration, and the excited state mainly by the D^+A^- configuration, excitation may be considered as involving the *transfer of an electron from D to A*. From the diagram it is apparent that the frequency of the charge-transfer band will depend on the energy gap ΔE. This means that complexes derived from good donor–acceptor pairs will absorb at relatively low frequencies. Accordingly, a linear correlation between the frequency of the charge-transfer band and the ionization potential of the donor (or electron affinity of the acceptor) is often observed. Several such correlations are illustrated in Fig. 7.3.

3. Equation (7.6) suggests that a given charge-transfer complex will be most stable when the value of B, the interaction parameter, is largest (the value of ΔE is predetermined by the choice of donor and acceptor). Since the value of B is related to the degree of overlap between the donor HOMO and the acceptor LUMO [see Eqs. (3.11) and (3.12)], the donor and acceptor should orient in such a way so as to optimize this HOMO–LUMO overlap. Crystallographic studies indeed confirm this expectation. For example, the complex between acetone and bromine, illustrated in **7.1**, shows an orientation that facilitates the overlap between an oxygen lone pair and the bromine σ^* orbital (the acceptor LUMO).[3] Complexes between aromatic derivatives invariably are sandwich-type structures, where the plane of one ring system faces the plane of the other. It is in this orientation that maximum overlap between the π_{HOMO} of the donor system and the π^*_{LUMO} of the acceptor system can occur.

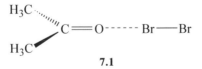

7.1

4. Despite the term charge-transfer complex, which implies there is substantial charge transfer in the ground state of the complex, the opposite is generally true: in most cases there is little charge transfer in the ground state. According to the Mulliken theory, the ground state is described primarily by the DA configuration, with D^+A^- contributing to just a small extent [Eq. (7.4)].

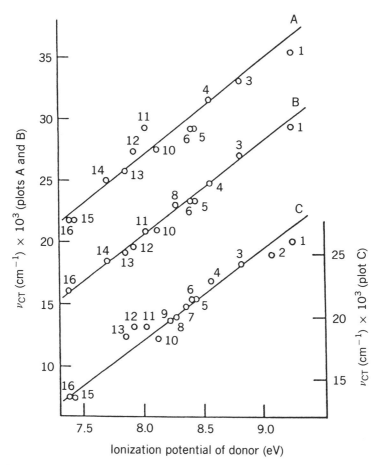

Fig. 7.3. Plot of charge-transfer (CT) band frequency ν_{CT} for a set of charge-transfer complexes with different aromatic donors, against the ionization potential of the donor molecule. The three lines show correlations for three acceptor molecules: 1,3,5-trinitrobenzene (line A), chloranil (line B), and tetracyanoethylene (line C). Donors are: (1) benzene, (2) chlorobenzene, (3) toluene, (4) *m*-xylene, (5) *p*-xylene, (6) mesitylene, (7) styrene, (8) biphenyl, (9) anisole, (10) naphthalene, (11) durene, (12) pentamethylbenzene, (13) hexamethylbenzene, (14) aniline, (15) azulene, (16) anthracene. [Reprinted with permission from Fig. 3.4 in R. Foster, *Organic Charge-Transfer Complexes*, Academic Press, NY, 1969. Copyright © Academic Press, 1969.

Numerically, the degree of charge transfer is determined by the mixing parameter λ ($\lambda < 1$). It is only in the excited state that the degree of charge transfer is substantial, since here the charge-transfer configuration dominates the excited-state wave function [Eq. (7.5)].

The above suppositions have been confirmed experimentally. Dipole moment studies confirm that the contribution of D^+A^- to the ground-state electronic structure is minor. For example, for complexes of aromatic compounds (e.g., durene, hexamethylbenzene, and naphthalene) with tetracyanoethylene, the extent of charge transfer has been estimated at about 6–7%, though for σ donors, such as amines, with iodine as an acceptor, the contribution of D^+A^- has been estimated as being as large as 35%. The charge-transfer character of the excited state has been confirmed in laser flash photolysis studies in which the excited charge-transfer state undergoes some chemical reaction. For example, photoexcitation of the anthracene–tetranitromethane complex, is found to generate the anthracene radical cation–tetranitromethane radical anion complex:

$$\text{Anthracene} \;//\; C(NO_2)_4 \xrightarrow{h\nu} [\text{Anthracene}]^{+\bullet}\,[C(NO_2)_4]^{-\bullet} \qquad (7.7)$$

Since the tetranitromethane radical anion within the excited-state ion-pair rapidly dissociates

$$[C(NO_2)_4]^{-\bullet} \xrightarrow{<3ps} C(NO_2)_3^{-} \;+\; NO_2^{\bullet} \qquad (7.8)$$

back-electron transfer to regenerate neutral reactants is avoided. Thus the observed products result from the combination of the anthracene radical ion with $C(NO_2)_3^{-}$ and $NO_2^{\bullet}$ (see Section 9.8).[4]

7.4 ELECTROCHEMICAL POTENTIALS

We stated that ionization potentials and electron affinities are primary physical parameters that govern the reactivity of a chemical species. Then, in principle, if one considers reactions in solution, *solution ionization potentials* and *solution electron affinities* would be required. In practice these parameters are not experimentally available, but as discussed in Chapter 8, they may be estimated from the gas-phase parameters and solvation energies. An alternative approach is to utilize standard electrochemical potentials, since in effect, these provide a measure of the ability of a species to give up, or accept, an electron *in solution.*

Consider the reaction

$$D + A \rightarrow D^+ + A^- \qquad (7.9)$$

This reaction is constructed from two half-reactions

$$A + e^- \rightarrow A^- \qquad (7.10)$$

$$D \rightarrow D^+ + e^- \qquad (7.11)$$

whose individual driving force can be measured electrochemically by placing a cell with an A/A^- (or D/D^+) electrode opposite a normal hydrogen electrode, whose potential is arbitrarily designated zero. When all the components of the system are in their standard state, the electromotive force of such a cell is then termed the standard redox potential of the A/A^- (or D^+/D) couple, measured in volts (V), and designated E°_{A/A^-} (or $E^{\circ}_{D^+/D}$). A selection of redox potentials for aromatic substrates taken from Eberson's recent monograph[5] is shown in Table 7.7. (A more extensive list, which incorporates indirectly determined and provisional E° values, can be found in Ref. 5.)

By convention the oxidant appears on the left-hand side of the equation. Also, half-reactions with more powerful oxidants than the standard hydrogen electrode are given *positive* E° values, while those with weaker oxidants are given *negative* E° values. Thus, for example, the redox potential for the benzene radical cation–benzene system is a large *positive* number (~ 3 V) since the benzene cation radical is a powerful oxidant while the redox potential for the benzene–benzene radical anion is a large *negative* number (~ -3 V) since benzene is a very weak oxidant.

Since the redox potential is a thermodynamic parameter it can be related to the equilibrium constants of the redox reactions. Thus the equilibrium constant for the redox reaction in Eq. (7.9) depends on the difference in the standard redox potentials, $E^{\circ}_{D^+/D}$ and E°_{A/A^-}

$$K = \exp \{F(E^{\circ}_{A/A^-} - E^{\circ}_{D^+/D})/RT\} \tag{7.12}$$

TABLE 7.7. The E° Values for Redox Couples of Selected Aromatic Compounds[a]

Reductant	Solvent[b]	E° (V)
Benzene	CH_3CN	3.03
Toluene	CH_3CN	2.61
p-Xylene	CF_3COOH	2.30
Naphthalene	CH_3CN	2.08
Anthracene	CH_3CN	1.61
Perylene	CH_3CN	1.30
[Tetracyanoethylene]$^-$	CH_3CN	0.48
[1,4-Dinitrobenzene]$^-$	DMF	-0.30
[Nitrobenzene]$^-$	DMF	-0.84
[Perylene]$^-$	CH_3CN	-1.44
[Anthracene]$^-$	CH_3CN	-1.74
[Naphthalene]$^-$	CH_3CN	-2.26
[Chlorobenzene]$^-$	DMF	-2.54
[Benzene]$^-$	1,2-Dimethoxyethane	-3.2

[a]Data taken from compilation in L. Eberson, *Electron Transfer Reactions in Organic Chemistry*, Springer-Verlag, Berlin, 1987, Table 1.
[b]*N,N*-Dimethylformamide is DMF.

TABLE 7.8. Vertical and Adiabatic Triplet Energies for Selected Alkenes

Molecule	Vertical E_T^a (kcal mol^{-1})	Adiabatic E_T^b (kcal mol^{-1})
Ethylene	99.6 (4.32)	84.0 (3.64)
Propene	98.7 (4.28)	
2-Methylpropene	97.3 (4.22)	80.5 (3.49)
trans-2-Butene	97.8 (4.24)	
Tetramethylethylene	94.5 (4.10)	75.8 (3.29)
Benzene		83.8 (3.63)
Toluene		82.1 (3.56)
Naphthalene		60.5 (2.62)
Cyclopentene		79.4 (3.44)
Cyclohexene		81.2 (3.52)
Styrene		64.9 (2.81)
trans-Stilbene		51.0 (2.21)

[a]Values from W. M. Flicker, O. A. Mosher, and A. Kupperman, *Chem. Phys. Lett.* **36**, 56 (1975). Values in electronvolts are shown in parentheses.
[b]Values from T. Q. Ni, R. A. Caldwell, and L. A. Melton, *J. Am. Chem. Soc.* **111**, 457 (1989). Values in electronvolts are shown in parentheses.

and the free energy of reaction is given by

$$\Delta G^\circ = F(E_{A/A^-}^\circ - E_{D^+/D}^\circ) \qquad (7.13)$$

where F is the Faraday constant (equal to the charge on a mole of electrons).

7.5 SINGLET–TRIPLET EXCITATION ENERGIES

We noted that adding an electron to a molecule or removing an electron from a molecule are fundamental processes that are intimately connected to reactivity. We saw, from the discussion of barrier formation in Section 4.4, that singlet–triplet excitation is also an important process; in electronic terms a range of reactions may be viewed as the *uncoupling* of a bond pair so as to facilitate the formation of a new bond pair. This was demonstrated for the model bond exchange process $H_2 + D_2 \rightarrow 2HD$, but applies to a range of basic organic reactions, primarily nonionic ones (pericyclic and radical reactions are the most important of these). Since the state that best describes the uncoupling of a bond pair is the triplet state, the ability to estimate singlet–triplet gaps is of value in assessing barrier heights in certain systems.

Vertical and adiabatic triplet energies for a selection of alkenes are listed in Table 7.8. As would be expected, vertical values (where the geometry of the system remains unchanged) are larger than adiabatic ones (where the system is allowed to relax). Note also that triplet energies mirror the π-bond strengths; the weaker the π-bond strength, the smaller the singlet–triplet excitation.

REFERENCES

1. R. S. Mulliken, *J. Am. Chem. Soc.* **74,** 811 (1952); R. S. Mulliken and W. B. Person, *Molecular Complexes, A Lecture and Reprint Volume*, Wiley, New York, 1969.

2. For a monograph on charge-transfer complexes, see R. Foster, *Organic Charge-Transfer Complexes*, Academic, New York, 1969.

3. O. Hassel and K. O. Stromme, *Nature* (London) **182,** 1155 (1958).

4. J. M. Masnovi and J. K. Kochi, *J. Am. Chem. Soc.* **107,** 7880 (1985).

5. L. Eberson, *Electron Transfer Reactions in Organic Chemistry*, Springer-Verlag, Berlin, 1987.

8

SOLVATION AND SOLVENT EFFECTS

8.1 IMPORTANCE OF SOLVENT

Due to the large advances in gas-phase chemistry over the last 20 years or so, it has become possible to compare reactivity in the gas phase with that in solution. The most obvious finding has been that the absolute magnitude of solvent effects on reactivity can be enormous; gas-phase and solution-phase reactivities are often dramatically different, not only in an absolute sense, but frequently in a relative sense as well. These studies have made it clear that the role of the solvent in governing a chemical reaction is far from passive. Given the fact that most chemical reactions are carried out in solution, it follows that a proper understanding of solvent effects is an essential component of any model of chemical reactivity.

Consider a simple nucleophilic substitution reaction: the identity exchange reaction

$$Cl^- + CH_3-Cl \rightarrow Cl-CH_3 + Cl^- \tag{8.1}$$

As is evident from Fig. 8.1, the energy profiles for this reaction in solution and in the gas phase differ in almost every respect. In the gas phase (illustrated by the lowest energy curve), the reactants come together initially in an energy releasing process to form an ion–molecule complex, $[Cl---CH_3Cl]^-$. This complex then overcomes an activation barrier to form a second ion–molecule complex (of the products) before dissociation to separated products. In aqueous solution, however (unbroken line in Fig. 8.1), no well-defined intermediate is formed along the reaction pathway so that the main feature is a large single barrier that separates solvated reactants from solvated products. What is par-

196

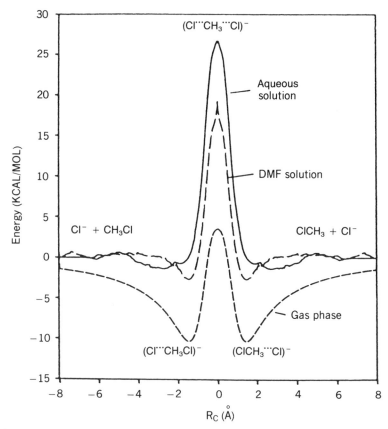

Fig. 8.1. Calculated energy profiles for the identity reaction of $Cl^- + CH_3Cl$ in the gas phase (lowest curve), in N,N-dimethylformamide (DMF) solution (middle curve), and in aqueous solution (top curve). [Reprinted by permission from J. Chandrasekhar and W. L. Jorgensen, *J. Am. Chem. Soc.* **107**, 2974 (1985). Copyright © American Chemical Society, 1985.]

ticularly striking though is that the barrier height in aqueous solution (26.6 kcal mol^{-1})[1] is some 10 times greater (!) than the overall gas-phase barrier (i.e., the barrier between separated reactants and the transition state), which is just 2.5 kcal mol^{-1}.[2] This barrier height difference of some 24 kcal mol^{-1} means *that reactivity in the gas phase is almost 18 orders of magnitude greater than in aqueous solution!* [Recall that from the equation $\Delta\Delta G^{\ddagger} = -RT \ln(k_1/k_2)$, it is simple to show that a change of 1.36 kcal mol^{-1} in the free energy of activation leads to a 10-fold change in the rate constant at 25°C. So an increase in the free energy barrier of 24 kcal mol^{-1}, would contribute to a rate constant increase of $24/1.36 = 17.6$ orders of magnitude.) In DMF solution (middle line in Fig. 8.1), the behavior is intermediate between that in the gas phase and in aqueous solution. Clearly solvent effects on reactivity can be very substantial.

Consider another nucleophilic substitution reaction

$$H_3N: + CH_3 - Br \rightarrow H_3N^+ - CH_3 + Br^- \qquad (8.2)$$

In this case the effect of solvent on the barrier height is the reverse! Here the gas-phase barrier is extremely high, while the solution reaction, named after its discoverer Menshutkin,[3] is facile and has been well known for a century. The dissociation of *tert*-butyl chloride into the *tert*-butyl cation and Cl^- provides another example of the same effect; in the gas phase this reaction is endothermic by about 150 kcal mol^{-1} while in water it is endothermic by just 20 kcal mol^{-1}, again demonstrating the enormous energetic effect the solvent can have on a chemical process.

How can we understand these huge and very different solvent effects on reactivity? The overall problem of solvent effects is very complex, and much of it still needs to be unraveled. The complexity arises because in the limit, each solvent molecule in the vicinity of the reactants may be viewed as part of the reaction in question. In principle, therefore, the coordinates of the solvent molecules are part of the reaction coordinates of the system. This means that a relatively simple reaction with just a few degrees of freedom becomes a multidimensional problem with an overwhelming number of reaction parameters. Such an approach requires huge amounts of computing power, and it is only recently, with the advent of fast computers and sophisticated statistical mechanical simulation programs, that attempts to treat dynamic solvent effects have been able to successfully reproduce reactivity in solution.[4]

While the recent successes of these quantitative models are encouraging, it is quite apparent that simpler models that provide qualitative insight in this question are needed as well. In view of the inherent complexity of the problem, the initial impression might be that the problem is intractable at a qualitative or simple quantitative level. However, on closer examination we find the situation is more promising than that; it turns out that many reactivity trends in the gas phase *are* reproduced in solution (though not all), so the solvation problem is not as intractable as it might initially appear. In fact, a number of simple models that deal with the role of solvent have been developed and found to be very useful. This chapter will try to look at some of the more common approaches.

8.2 HYDRATION ENERGIES OF MOLECULES AND IONS

Before we discuss solvent effects on reactivity in any detail, we need to first survey solvation energies of neutral and charged species and, by extension, the way solvation phenomena affect reaction equilibria. These data will serve as important elements in a model of solvation effects on reactivity.

Free energies of hydration of neutral polar molecules are generally small. From group activity coefficients[5] it can be estimated that the aqueous solvation energies of alcohols, ethers, amines, and carboxylic acids are all approximately

TABLE 8.1. Hydration Energies of Ions $\Delta G_{hyd}{}^a$

Ion	ΔG_{hyd} (kcal mol^{-1})	Ion	ΔG_{hyd} (kcal mol^{-1})
$PhCH_2^-$	-57	NH_4^+	-77
PH_2^-	-65	$CH_3NH_3^+$	-68
NH_2^-	-93	$(CH_3)_2NH_2^+$	-61
HSe^-	-65	$(CH_3)_3NH^+$	-57
$CH_3COCH_2^-$	-79	$PhNH_3^+$	-66
$CH_3CH_2CH_2S^-$	-74	PH_4^+	-71
PhS^-	-65	H_3O^+	-102
NO_2^-	-70	$CH_3OH_2^+$	-83
PhO^-	-70	H_3S^+	-85
HS^-	-74	$CH_3SH_2^+$	-72
CH_3O^-	-93	H^+	-259.5
I^-	-61		
Br^-	-70		
Cl^-	-75		
F^-	-105		
CN^-	-75		
CH_3COO^-	-75		
$CF_3CH_2O^-$	-81		
N_3^-	-72		
HO^-	-104		
HCC^-	-71		

^{a}Data taken from R. G. Pearson, *J. Am. Chem. Soc.* **108**, 6109 (1986).

-3 ± 2 kcal mol^{-1}. In contrast ionic hydration energies, as would be expected, are substantially larger. Hydration energies of cations and anions have been estimated using thermodynamic cycles and some results are listed in Table 8.1. Values typically lie in the range -50 to -100 kcal mol^{-1}, though the H^+ ion value of -259.5 is especially high. We will subsequently see that ionic solvation energies are an important contributor to the barrier in solution of reactions in which ions are involved.

8.3 SOLVENT EFFECTS ON ACID–BASE REACTIONS

Acid–base reactions are among the most studied in chemistry. The wealth of data both in the gas phase and in solution provides valuable insight into solvent effects on the thermodynamics of the two simple dissociation processes

$$HX \rightarrow H^+ + X^- \tag{8.3}$$

$$HY^+ \rightarrow H^+ + Y \tag{8.4}$$

Table 8.2 lists the free energy of dissociation in the gas phase $\Delta G°(g)$, in DMSO $\Delta G°(sol)$, and in aqueous solution $\Delta G°(aq)$, for a range of organic and

TABLE 8.2. Free Energies of Acid Dissociation in the Gas Phase, $\Delta G_{(g)}^{\circ}$, in Dimethylsulfoxide, $\Delta G_{(sol)}^{\circ}$, and in Aqueous Solution, $\Delta G_{(aq)}^{\circ}$

Compound	$\Delta G_{(g)}^{\circ}$ [a] (kcal mol^{-1})	$\Delta G_{(sol)}^{\circ}$ [a] (kcal mol^{-1})	$\Delta G_{(aq)}^{\circ}$ [b] (kcal mol^{-1})
CH_4	408.5	≈ 76	≈ 68
H_2O	384.1	43.0	21.5
CH_3OH	374.0	39.7	21.2[c]
$PhCH_3$	373.7	58.8	56.0
EtOH	370.8	40.8	21.7[c]
$(CH_3)_3COH$	368.0	44.0	24.6
HF	365.7	≈ 20.5	4.4
CH_3CN	365.2	42.8	≈ 34
CF_3CH_2OH	354.1	31.2	16.9
Ph_3CH	352.8	41.9	42
CH_3NO_2	349.7	23.6	13.6
Fluorene	344.0	31.0	31.4
HCN	343.8	17.7	12.4
PhOH	342.3	24.7	13.6
CH_3CO_2H	341.1	16.4	6.5
p-Nitroaniline	336.2	28.5	25.1
PhCOOH	333.0	15.2	5.8
$CH_2(CN)_2$	328.9	15.4	15.0
HBr	318.2	1.1	-12.3
p-Nitrophenol	317.3	14.9	9.8
CF_3COOH	316.3	4.9	0.3
$(CH_3)_3NH^+$	221[d]		13.3
$(CH_3)_2NH_2^+$	217[d]		14.6
$CH_3NH_3^+$	210[d]		14.5
NH_4^+	198[d]		12.6

[a] $\Delta G_{(g)}^{\circ}$ and $\Delta G_{(sol)}^{\circ}$ defined as the free energy of Eq.(8.3) [or Eq. (8.4)] in the gas phase and in dimethylsulfoxide (DMSO), respectively. Values derived from data in R. W. Taft and F. G. Bordwell, *Acc. Chem. Res.* **2**, 463 (1988).
[b] $\Delta G_{(aq)}^{\circ}$ defined as the free energy of Eq. (8.3) [or Eq. (8.4)] in aqueous solution. Values calculated from aqueous pK_a data and Eq. (8.6). Compilations of pK_a values may be found in E. P. Serjeant and B. Dempsey, *Ionisation Constants of Organic Acids in Aqueous Solution*, 2nd ed., Pergamon Press, Oxford, 1979; D. D. Perrin, *Dissociation Constants of Organic Bases in Aqueous Solution*, Butterworths, London, 1965, and Suppl., 1973.
[c] Values are uncertain; based on pK_a of 15.5 for CH_3OH and 15.9 for EtOH.
[d] Data taken from C. D. Ritchie. In *Solvent–Solute Interactions*, Vol. 2, J. F. Coetzee and C. D. Ritchie, Ed., Dekker, New York, 1976, Chapter 12.

inorganic acids. Solution ΔG° values and pK_a values are directly related to one another through the relationship

$$\Delta G^{\circ} = -2.303RT \log K_a \qquad (8.5)$$

which, at 25°C, simplifies to

$$\Delta G^{\circ} = 1.365 \, pK_a \qquad (8.6)$$

Thus the solution $\Delta G°$ data in Table 8.2 (expressed in kcal mol^{-1}) can be readily converted to solution pK_a values simply by dividing the data by 1.365.

The most obvious feature in Table 8.2 is that free energies of dissociation in the gas phase are substantially larger than those in solution; neutral acids in the gas phase have dissociation energies in excess of 300 kcal mol^{-1}, while solution values rarely exceed 50 kcal mol^{-1}. Solvation of the ions formed from the dissociation in solution explains this huge energetic difference. Note that H^+ solvation on its own, accounts for 259.5 kcal mol^{-1} of this difference. Solvation affects not only the *magnitude* of the dissociation free energy but the *order* as well and sometimes this effect is quite dramatic. The compounds in Table 8.2 are listed in order of increasing gas-phase acidity. However, perusal of the $\Delta G°$(aq) values shows the solution order to be quite different. For example, PhCH$_3$ is *more* acidic in the gas phase than H_2O (by 10.4 kcal mol^{-1}), but much *less* acidic in aqueous solution (by 34.5 kcal mol^{-1}, the equivalent of about 25 pK_a units). Carbon acids generally (e.g., fluorene, Ph$_3$CH, or CH$_3$CN) are surprisingly strong acids in the gas phase given their low acidity in aqueous solution. Thus Ph$_3$CH and trifluoroethanol have similar gas-phase acidity even though in aqueous solution, trifluoroethanol is more acidic by about 25 kcal mol^{-1} (equivalent to about 18 pK_a units!).

Why is the order of gas-phase and solution acidity different? The extent of charge dispersal in the conjugate base anion is the key to understanding the different gas-phase and solution acidity orders. Inspection of the data in Table 8.1 reveals that *localized ions are more strongly solvated than delocalized ions.* For example, the hydration energy of the phenoxide ion (70 kcal mol^{-1}) is smaller than that of the methoxide ion (93 kcal mol^{-1}), since part of the negative charge in phenoxide is delocalized over the benzene ring. The solvation energy of the benzyl anion is even smaller (57 kcal mol^{-1}); replacement of the electronegative O atom by the less electronegative CH$_2$ group increases the degree of charge delocalization into the ring even more.

What these differences in solvation energies mean is *that neutral acids with highly delocalized conjugate base anions will be relatively strong acids in the gas phase, while those whose conjugate base anions are strongly localized will be relatively strong acids in aqueous solution.* In effect, delocalized anions can be thought of as being solvated *intramolecularly*—the charge is already dispersed over several atoms, and thus less dependent on *intermolecular* solvation. Clearly, gas-phase and solution acidity will not correlate for a range of acids whose conjugate bases are delocalized to different degrees.

8.3.1   Effect of Alkyl Groups on Acidity and Basicity

The introduction of methyl substituents into the water molecule leads to an increase in ROH gas-phase acidity. Thus acidity increases along the series:

$$H_2O < CH_3OH < CH_3CH_2OH < (CH_3)_2CHOH < (CH_3)_3COH$$

Remarkably, this ordering is largely reversed in solution with H_2O being *more* acidic than $(CH_3)_3COH$, though the precise acidities of the simple alcohols in water are somewhat uncertain.

In the gas phase the ordering appears to be governed by the ability of the alkyl group to stabilize the negative charge on oxygen. The larger the group, the more effective the stabilization. This is another example of what we termed earlier as intramolecular solvation; as the size of the alkyl group increases, the ability of the group to disperse the negative charge on O^- increases as well. Thus alkyl groups may act as electron acceptors when adjacent to negative charge, or as electron donors when adjacent to positive charge, i.e. alkyl groups can stabilize both cations and anions.

In solution, the acidity of the alcohols is strongly influenced by solvation effects; the larger the solvation energy of the conjugate base anion, the greater the acidity of the parent alcohol. Since it is the smaller anions (e.g., HO^- or CH_3O^-) that have the larger solvation energy, the effect of solvation on the acidity order operates in reverse to the intrinsic molecular effect. Overall, the solvation effect turns out to be larger than the structural effect, and therefore the solution acidity order is governed by solvation energies. Hence, the acidity order in solution is almost the reverse of the one observed in the gas phase.

A similar effect is observed for alkyl amines (Table 8.2). In the gas phase, the acidity of the protonated amines

$$R_3NH^+ \rightarrow R_3N + H^+ \tag{8.7}$$

increases in the order

$$(CH_3)_3NH^+ < (CH_3)_2NH_2^+ < CH_3NH_3^+ < NH_4^+$$

This increase reflects the ability of the methyl groups to stabilize the positive charge in the cation, and thereby *increase* the free energy of deprotonation, $\Delta G_{(g)}^\circ$. However, in solution no clear trend is observed due to partial cancellation of the intrinsic substituent effect by a solvent effect.

Free energies of acid dissociation in DMSO listed in Table 8.2 are in some cases close in value to those in water, but in other cases they differ significantly. The reason for the different behavior of H_2O and DMSO—a so-called dipolar aprotic solvent—is described in Section 8.6.

8.4 SOLVENT POLARITY AND THE HUGHES–INGOLD MODEL

Let us now consider the problem of solvent effects on reactivity. This problem is a more difficult one than understanding solvent effects on reaction thermodynamics, since it requires an understanding of how the solvent will affect that transient structure: the transition state.

The simplest model of solvent effects on reactivity is based on the *polarity* of the solvent, as measured by its dielectric constant ϵ. Using this approach the solvent is treated as a continuous medium and the specific molecular structure of the solvent molecules is ignored. The model rests on the physical observation that the transfer of a charge or a dipole from a vacuum ($\epsilon = 1$) to a medium of dielectric constant ϵ, leads to a reduction in the free energy of the charge or dipole; the greater the dielectric constant, the larger the stabilizing effect. For example, the transfer of a dipole from a vacuum ($\epsilon = 1$) to a medium of dielectric constant ϵ is given by

$$\Delta G = -\mu^2/r^3\{(\epsilon - 1)/(2\epsilon + 1)\} \qquad (8.8)$$

where μ is the molecular dipole moment, and r is the radius of the molecule, which is assumed to be spherical. It was on this basis that Hughes and Ingold[6] proposed that *reactions in which charge is generated, or charge separation is increased, will be assisted by polar solvents, and conversely, reactions in which charge is annihilated, or charge separation is reduced, will be inhibited by polar solvents.*

The effect of solvent on the Menshutkin reaction [Eq. (8.2)] or on the dissociation of *tert*-butyl chloride is now clearer. Both of these processes involve charge separation, so that formation of ions will be greatly assisted by solvent. Furthermore, the more polar the solvent, the greater the effect will be. The effect of solvent polarity on a number of different reaction types is summarized in Table 8.3. Solvents with low dielectric constant ($\epsilon < 15$) are classified as *nonpolar*, while those with high dielectric constant ($\epsilon > 15$) are classified as *polar*. The dielectric constants for some common solvents are listed in Table 8.4.

From Table 8.3 it can be seen that reactions in which charge is generated, are *strongly assisted* by a polar solvent, while reactions in which charge is annihilated, are *strongly retarded* by a polar solvent. Reactions in which charge is dispersed (i.e., the charge is distributed over more atoms) are *slightly retarded* by a polar solvent.

Note that the reactivity pattern described in Table 8.3 for charge separation or charge annihilation reactions is another example of the general link between rates and equilibria (Section 6.5). Changes in solvent polarity for such reactions

TABLE 8.3. Effect of Solvent Polarity on Reaction Rate

Reaction Type	Nature of Charge Type	Effect of Increase of Solvent Polarity on Reaction Rate
$R-X \rightarrow R^{\delta+} \cdots X^{\delta-}$	Separation	Large increase
$Y + R-X \rightarrow Y^{\delta+} \cdots R \cdots X^{\delta-}$	Separation	Large increase
$R-X^+ \rightarrow R^{\delta+} \cdots X^{\delta+}$	Dispersion	Small decrease
$Y^- + R-X \rightarrow Y^{\delta-} \cdots R \cdots X^{\delta-}$	Dispersion	Small decrease
$Y^- + R-X^+ \rightarrow Y^{\delta-} \cdots R \cdots X^{\delta+}$	Annihilation	Large decrease

TABLE 8.4. Polarity Data for Some Common Solvents

Solvent	ϵ^a	Y^b	Z^c	E_T^d
Water	78.5	3.49	94.6	63.1
Methanol	32.7	−1.09	83.6	55.5
Ethanol	24.6	−2.03	79.6	51.9
80% Ethanol–water	67.0	0.00	84.8	53.6
2-Propanol	18.3	−2.73	76.3	48.6
tert-Butanol	12.5	−3.26	71.3	43.9
Formic acid	57.9	2.05		
Acetic acid	6.15	−1.68	79.2	
Trifluoroacetic acid	8.32	≈4.5	≈88	
Formamide	111.0	0.6	83.2	56.6
Acetonitrile	37.5		71.3	46.0
Dimethylsulfoxide	46.7		71.1	45.0
N,N-Dimethylformamide	36.7		68.5	43.8
Acetone	20.7		65.7	42.2
Pyridine	12.4		64.0	40.2
Tetrahydrofuran	7.58			37.4
Hexamethylphosphoramide	30		62.8	40.9
Chloroform	4.64		63.2	
Carbon tetrachloride	2.24			32.5
Benzene	2.28		54	34.5
Hexane	1.88			30.9

[a]Most dielectric constant data are taken from J. A. Riddick and W. B. Bunger. In *Organic Solvents*, 3rd ed., Vol. II, A. Weissberger, Ed., *Techniques of Chemistry*, Wiley-Interscience, New York, 1970.
[b]From E. Grunwald and S. Winstein, *J. Am. Chem. Soc.* **70**, 846 (1948).
[c]From E. M. Kosower, *J. Am. Chem. Soc.* **80**, 3253 (1958).
[d]From K. Dimroth, C. Reichardt, T. Siepman, and F. Bohlmann, *Ann. Chem.* **661**, 1 (1963); C. Reichardt, *Angew. Chem. Int. Ed. Engl.* **4**, 29 (1965).

affect the thermodynamics of the reaction dramatically. Consequently, the large rate changes observed in these reactions are simply a kinetic manifestation of the large change in their thermodynamics. Attempting to understand the way in which solvent affects reactivity in those cases where solvent changes do *not* affect reaction thermodynamics (e.g., in identity reactions) is much more difficult.

8.4.1 Empirical Solvent Polarity Scales

The classification of different reaction types shown in Table 8.3 demonstrates the importance of solvent polarity. However, if one attempts to actually correlate reactivity with solvent polarity, as measured by the appropriate function of dielectric constant [Eq. (8.8)], substantial scatter is observed. Clearly other factors are superimposed on the general polarity function. In order to overcome this difficulty in correlating data, an *empirical* measure of polarity was proposed by Grunwald and Winstein,[7] based on a model reaction of charge separation. The model reaction chosen was the rate of solvolysis of *tert*-butyl chloride. This reaction was chosen since the rate-determining step

$$(CH_3)_3C-Cl \rightarrow (CH_3)_3C^{\delta+} \cdots Cl^{\delta-} \qquad (8.9)$$

involves the S_N1 ionization of the reactant in a charge-separation process. The rate constant for such a reaction would therefore be expected to reflect in a direct way the ability of a particular solvent to assist in the ionization process. Accordingly, a solvent polarity measure Y was defined by

$$\log (k/k_0) = Y \qquad (8.10)$$

where k_0 is the rate constant for the solvolysis of *tert*-butyl chloride in 80% ethanol–water (the reference solvent) at 25°C, and k is the corresponding rate constant in any other solvent.[7] From Eq. (8.10) it follows that the Y value for 80% ethanol–water is defined as zero. A positive Y value for a solvent indicates it to be *more* polar than 80% ethanol–water, while a negative Y value indicates the solvent to be *less* polar than 80% ethanol–water. The Y values for representative solvents are shown in Table 8.4. For example, the Y value of 3.49 for water means that the rate of solvolysis of *tert*-butyl chloride in water is almost four orders of magnitude faster than in 80% ethanol–water. So in contrast to ϵ, which is a more theoretically based measure of solvent polarity, Y values are intended to provide a direct *empirical* measure of that property.

Using the rate of solvolysis of *tert*-butyl chloride as a measure of solvent polarity is not practical for many solvents. The solvents may not react in a simple solvolysis reaction, or alternatively, may be too reactive for rate constants to be conveniently measured. To overcome this problem, solvent polarity scales based on a spectroscopic process have been proposed. The most important of these are the Z and E_T scales. Both scales are spectroscopically derived and are based on the energy of a charge-transfer transition. The Z scale is based on the process:[8]

$$(8.11)$$

Since the ground-state ion-pair is highly polar, its energy is strongly dependent on solvent polarity, while the charge-transfer product, being neutral, is relatively unaffected. Consequently, the effect of solvent on the energy of the charge-transfer transition is a useful measure of solvent polarity. The magnitude of the Z value is just the energy of the charge-transfer transition measured in kilocalories per mole. The E_T scale is based on the same principle, but the energy of the charge-transfer transition is measured for the pyridinium zwitterion (**8.1**) in an *intramolecular* process.[9] Both Z and E_T values for some solvents are listed in Table 8.4.

8.1

8.4.2   Winstein–Grunwald Equation

A natural extension of the equation defining Y values, Eq. (8.10), is to incorporate a *solvent reaction parameter m*, so that the sensitivity of *different* reactions to changes in solvent polarity may be gauged. The resulting equation, termed the Winstein–Grunwald equation,[7] is written as

$$\log (k/k_0) = mY \qquad (8.12)$$

where k_0 is the rate constant for a particular reaction in 80% ethanol–water, k is the corresponding rate constant in some solvent, and m defines the *sensitivity* of the particular reaction to changes in solvent polarity. From its logarithmic form it is clear that Eq. (8.12) is a linear free energy relationship for solvent effects, and in that sense it is analogous to the Hammett equation—the linear free energy relationship for substituent effects. Thus the m value is the analog of the Hammett reaction parameter ρ, while Y is the analog of the Hammett substituent parameter σ. For the reference reaction (*tert*-butyl chloride solvolysis) the m value is, by definition, equal to 1. Other reactions may have m values that are larger or smaller than 1 and this depends primarily on the degree of charge separation in the transition state of the reaction in question.

The scope of the Winstein–Grunwald equation is more limited than that of the Hammett equation. Establishing an empirical scale of solvent polarity does not solve the question of solvent effect on reactivity because even the *order* of solvent reactivity (as defined by the Y scale) is not invariant. Solvent effects on different reactions are found to differ widely. For example, whereas the solvolysis of *tert*-butyl chloride in trifluoroacetic acid is almost seven orders of magnitude *faster* than in ethanol (as reflected by the relative Y values of these two solvents), the solvolysis of ethyl tosylate in trifluoroacetic acid is two orders of magnitude *slower* than in ethanol. Clearly, these two reactions respond to solvent effects in quite different ways. In order to understand reactivity inversions of this kind, a more detailed understanding of *specific solvent–solute interactions* is required.

8.5 SOLVENT ELECTROPHILICITY AND NUCLEOPHILICITY

As the solvent becomes more polar, its ability to solvate polar and ionic solutes increases. However, a model that considers the solvent to be a continuous dielectric medium is just a crude approximation. In reality solvent molecules undergo *specific interactions* with solute molecules—for ionic and polar solutes the most important of these are charge–dipole and dipole–dipole interactions. These interactions are of two main types: *nucleophilic* and *electrophilic*. In a nucleophilic interaction, a point of positive charge in the solute is stabilized by interacting with the negative end of the solvent dipoles as in **8.2**, while in an electrophilic interaction, a point of negative charge in the solute is stabilized by the positive end of the solvent dipoles as in **8.3**.

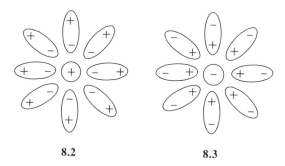

8.2   **8.3**

So it is not just the polarity of a solvent, as measured by ϵ, that determines the solvent effect on reactivity, but also its ability to provide specific nucleophilic and electrophilic solvation. These three properties (polarity, nucleophilicity, and electrophilicity) enable solvents to be usefully classified according to the following scheme:

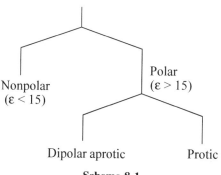

Scheme 8.1

Solvents are primarily classified into polar and nonpolar categories ($\epsilon = 15$ is the somewhat arbitrary value that divides these two groups). A further classification of great importance for the polar solvents is into *protic* and *aprotic* types. Protic solvents (e.g., water, ethanol, or formic acid) possess a relatively acidic hydrogen, so these molecules are able to effectively stabilize negative charge through a hydrogen bond, and positive charge through some nucleophilic site in the molecule. In other words, protic solvents possess both nucleophilic *and* electrophilic character.

Consider the solvation of Na^+ and Cl^- ions by a water molecule as an example of solvation by a protic solvent. The positive charge on Na^+ can be stabilized by interacting with the negative end of the water dipole (the oxygen atom), as shown in **8.4**, while the negative charge on Cl^- can be stabilized by interacting with the positive end of the water dipole—one of the hydrogen atoms, as shown in **8.5**.

8.4   **8.5**

The high solvating power of trifluoroacetic acid, as reflected in a particularly large Y value (~ 4.5), may now be understood. Even though this solvent has a low dielectric constant ($\epsilon = 8.3$), much lower than that of ethanol ($\epsilon = 24.6$), its effectiveness as a solvent for *tert*-butyl chloride solvolysis is a remarkable six orders of magnitude *greater* than that of ethanol. The special ability of trifluoroacetic acid to catalyze an S_N1 reaction comes about from its very high acidity ($pK_a = 0.2$). This facilitates strong hydrogen bonding to the leaving group, as illustrated in **8.6**. Thus trifluoroacetic acid can be a very effective solvent for ionization processes due to its powerful *electrophilic* character.

8.6   **8.7**

Now consider the solvolysis of ethyl tosylate (EtOTs). The mechanism of this reaction, being S_N2 in nature, depends greatly on *nucleophilic* solvation. Solvolysis of ethyl tosylate cannot take place without substantial nucleophilic interaction of the solvent with the substrate in the transition state, depicted in

8.7. So for this reaction the nucleophilic solvent ethanol is a better solvent than the nonnucleophilic solvent trifluoroacetic acid, which is less effective in supplying the required nucleophilic assistance. In other words, solvent effects are highly reaction specific; a good solvent for one reaction may well be a poor solvent for a second reaction and vice versa.

The Winstein–Grunwald Y value was designed to quantify the ionizing power of the solvent. Actually, it was later discovered that even the supposedly "pure S_N1" substrate, tert-butyl chloride, undergoes some nucleophilic assistance in most solvents. This was discovered by comparing the solvent effect on tert-butyl chloride solvolysis with the solvolysis of 1-adamantyl derivatives (**8.8**). The ionization of 1-adamantyl derivatives *cannot* be nucleophilically assisted by solvent due to their cage structure; back-side approach of a nucleophile is precluded by the molecular architecture. A plot of log k for tert-butyl chloride solvolysis against log k for 1-adamantyl chloride solvolysis for a wide range of solvents is shown in Fig. 8.2.[10] The plot shows a good correlation for most solvents. However, the highly ionizing, nonnucleophilic fluorinated solvents fall substantially *below* the correlation line. In the fluorinated solvents, tert-butyl chloride appears to solvolyze more *slowly* than predicted by the correlation line. These results led to the conclusion that the ionization of tert-butyl chloride in the relatively nucleophilic water and alcohol solvents *is* assisted by some nucleophilic participation so that tert-butyl chloride was not an ideal substrate on which to model solvent ionizing ability.

8.8

An attempt to improve on the Winstein–Grunwald equation by incorporating an additional term to account for nucleophilic assistance has been proposed.[11] The improved equation

$$\log (k/k_0) = mY + lN \qquad (8.13)$$

incorporates two additional parameters: N, a solvent nucleophilicity parameter defined for solvents on the basis of the rate of nucleophilic substitution of methyl tosylate, the reference substrate; and l, a sensitivity parameter, that defines the sensitivity of any substrate to nucleophilic assistance. The parameter l is defined to be 1 for the reference substrate. The model compound chosen for assessing the solvent ionizing ability Y is 1-adamantyl chloride, rather than tert-butyl chloride, since the structure of the adamantyl system precludes *any* nucleophilic assistance. As before, an m value of 1 is defined for the reference substrate.

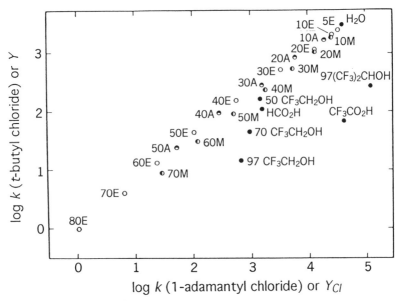

Fig. 8.2. Correlation of rate constants for solvolysis of 1-adamantyl chloride versus those for *tert*-butyl chloride in a wide range of solvent mixtures at 25°C. [Reprinted by permission from T. W. Bentley and G. E. Carter, *J. Am. Chem. Soc.* **104**, 5741 (1982). Copyright © American Chemical Society, 1982.]

Unfortunately, even incorporating these additional parameters does not fully resolve the problem of correlating rate data. It turns out that not only is there variability in the degree of nucleophilic assistance that different substrates enjoy, but the degree of electrophilic assistance is also substrate dependent. In other words the assumption that electrophilic assistance to ionization may be incorporated as a fixed component of the Y term proved to be unsatisfactory. Different leaving groups respond differently to solvent electrophilic assistance so that even the four parameter equation [Eq. (8.13)] cannot fully correlate available rate data. One way around this problem is to introduce different Y scales for different leaving groups but this approach of course detracts from the generality of the procedure. An alternative approach to overcoming this problem would be to incorporate additional parameters to reflect the variable electrophilicity requirements of different substrates. However, this would require a six-parameter equation to correlate rate data, and this ceases to be of practical value.

Therefore, we must conclude that the goal of being able to reliably estimate the rate of some reaction in a particular solvent, based on the rate of a reference reaction in a reference solvent, will remain out of reach. In seeking solvent correlations, we cannot hope to obtain a general solvent correlation equation, but must restrict ourselves to the correlation of either similar reaction or solvent types.

8.6 DIPOLAR APROTIC SOLVENTS

The class of polar *aprotic* solvents (or dipolar aprotic as they are commonly termed) is a special group of solvents with important solvating properties. This group is distinguished from the polar protic solvents by the absence of any acidic hydrogen atoms. Examples of some dipolar aprotic solvents are acetone, acetonitrile, dimethylformamide (DMF), dimethylsulfoxide (DMSO), and hexamethylphosphoramide. Whereas the polar protic solvents can stabilize both positive and negative charge, dipolar aprotic solvents can stabilize positive charge effectively, but are *less effective* in stabilizing negative charge. In these molecules the positive end of the molecular dipole tends to be located deeper within the molecular space so that the ability of the molecule to effectively interact with negative charges in the solute is reduced.

The different abilities of solvents to solvate ions may be quantified by measuring the free energy of transfer ΔG_{tr}, of ions from one solvent to another. Unfortunately, absolute single-ion data is impossible to come by; all ions necessarily are associated with a counterion so that any attempt to obtain single-ion data must rest on some assumption. An assumption that is commonly used is that the free energy of transfer of $Ph_4As^+Ph_4B^-$ between two solvents may be split equally between the two ions, Ph_4As^+ and Ph_4B^-.[12] Since these two ions have similar size and shape and since the charge is buried deep within the ion, it is assumed that solvation of the two ions is the same. Once single-ion free energy values for these ions are known, values for other ions can be obtained by comparing salts that share a common ion. This approach has been extensively employed by Parker[13] and illustrative free energy of transfer data for selected ions in several solvents are listed in Table 8.5. A positive sign of the ΔG_{tr} value tells us the free energy of transfer is endergonic, indicating that the initial solvent is the better solvent. With the exception of the unusually

TABLE 8.5. Free Energies of Transfer of Ions from Methanol to DMF, DMSO, and Water[a]

Ion	ΔG_{tr}(methanol → DMF)	ΔG_{tr}(methanol → DMSO)	ΔG_{tr}(methanol → water)
Cl^-	8.8	7.5	−3.4
Br^-	6.7	4.9	−2.9
I^-	3.5	1.8	−2.0
N_3^-	6.7	4.8	−2.9
CN^-	8.4		−2.4
OAc^-	12.5	8.8	−3.9
SCN^-	3.7	1.9	−1.2
Ag^+	−6.9	−11.2	−1.1
K^+	−5.0	−6.1	−2.0
Na^+	−5.3	−4.9	
Ph_4As^+	−3.7	−3.5	5.6

[a]Data in kilocalories per mole obtained from solvent activity coefficients in A. J. Parker, *Chem. Rev.* **69**, 1 (1969).

large Ph_4As^+ ion, free energy of transfer of ions from methanol to water (a more polar solvent) is exergonic. Of special interest, however, are the positive ΔG_{tr} values for anions (ranging between 3 and 12 kcal mol^{-1}) when transferred from a protic to an aprotic solvent. These positive values confirm that anions in dipolar aprotic solvents are poorly solvated (the effect is more pronounced in DMF than in DMSO). Cations, on the other hand, *are* effectively solvated in the dipolar aprotic solvents, as reflected by the negative values of ΔG_{tr}.

The special solvation properties of dipolar aprotic solvents also show up in the acid dissociation free energies listed in Table 8.2. In all cases free energy values in DMSO, $\Delta G°(sol)$, are either equal to or less than the corresponding values in aqueous solution, $\Delta G°(aq)$. For delocalized anions, such as those obtained from the acids Ph_3CH, fluorene, and $CH_2(CN)_2$, where solvation is less important, $\Delta G°(sol)$ and $\Delta G°(aq)$, are effectively the same. However, for more localized anions, where stabilization through hydrogen bonding increases, $\Delta G°(sol)$ values are *greater* than the corresponding $\Delta G°(aq)$ values; that is, it is more difficult to dissociate an acid, such as CH_3OH (which forms a localized anion) in DMSO than in water. This special quality of dipolar aprotic solvents of being able to effectively solvate cations, but not anions, has important chemical consequences, which we will now discuss.

The rate of nucleophilic reactions of anions (such as in S_N2, additions to carbonyls) are dramatically enhanced by 2–6 orders of magnitude when carried out in dipolar aprotic solvents. For example, the rate of reaction

$$N_3^- + CH_3Cl \rightarrow N_3CH_3 + Cl^- \tag{8.14}$$

is over 1 million times faster in DMF than in methanol.[13] This remarkable rate enhancement is attributed to the relatively weak solvation of anionic nucleophiles in dipolar aprotic solvents as reflected in their positive ΔG_{tr} values. In polar protic solvents, by contrast, the nucleophile is strongly solvated so that the reaction is relatively inhibited. What these results suggest is that *desolvation of the nucleophile is an important component of the barrier to nucleophilic substitution in solution.* (Recall that the addition of each 1.36 kcal mol^{-1} to the activation barrier contributes to a reduction in rate of a factor of 10. Hence, a positive free energy of transfer of 6 kcal mol^{-1}, would contribute to a rate constant *decrease* of over four orders of magnitude.)

A free energy diagram that illustrates the S_N2 barrier in a protic and aprotic medium is shown in Fig. 8.3. While both the reactants and the transition state are destabilized when transferred from a protic solvent (firm line) to an aprotic solvent (dashed line), it can be seen that the main reduction in the free energy barrier is due to the destabilization of the reactants, and this is due to weaker nucleophile solvation. This analysis is consistent with the negligible *gas-phase* barrier to substitution (discussed in Section 8.1) where, of course, solvation is totally absent.

It is possible to confirm that it is mainly the reduction in nucleophile solvation that is responsible for the large reactivity change in transferring from a

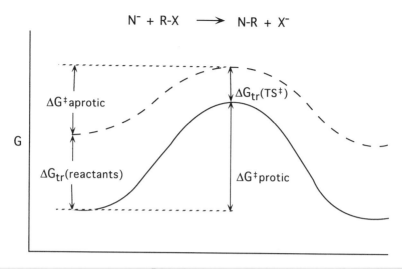

Fig. 8.3. Free energy diagram showing the effect of solvent on the energy profile of an anionic S_N2 reaction. Dashed line shows profile for reaction in a dipolar aprotic solvent; firm line shows profile for reaction in a protic solvent.

protic solvent to a dipolar aprotic one. This can be done by dissecting the solvent effect on the rate constant into its components—the solvent effect on the free energy of transfer of the reactants $\Delta G_{tr}(RX)$ and $\Delta G_{tr}(X^-)$, and on the free energy of transfer of the transition state $\Delta G_{tr}(TS^{\ddagger})$. Both $\Delta G_{tr}(RX)$ and $\Delta G_{tr}(X^-)$ are measured directly while $\Delta G_{tr}(TS^{\ddagger})$ is estimated from the rate constants. Let us see how this is done.

From transition state theory the rate constants in the two solvents $k_{aprotic}$ and k_{protic} are related to the difference in the free energy of activation in the two solvents $\Delta\Delta G^{\ddagger}$ by

$$\Delta\Delta G^{\ddagger} = -RT \ln k_{aprotic}/k_{protic} \tag{8.15}$$

where $\Delta\Delta G^{\ddagger}$ is defined by $\Delta G^{\ddagger}_{aprotic} - \Delta G^{\ddagger}_{protic}$.

From Fig. 8.3 it is easy to show that

$$\Delta\Delta G^{\ddagger} = \Delta G_{tr}(TS^{\ddagger}) - \Delta G_{tr}(RX) - \Delta G_{tr}(X^-) \tag{8.16}$$

So it is possible to obtain the value of $\Delta G_{tr}(TS^{\ddagger})$ from the free energy of transfer of the reactants $\Delta G_{tr}(RX)$ and $\Delta G_{tr}(X^-)$, and from the magnitude of $\Delta\Delta G^{\ddagger}$ as assessed from the rate ratio $k_{aprotic}/k_{protic}$ by using Eq. (8.15).

Parker has carried out this analysis for a large number of reactions and the approach is illustrated in Table 8.6 This table presents the free energy of transfer of the substrate, nucleophile, and transition state from methanol to

TABLE 8.6. Free Energy of Transfer ΔG_{tr} of Reactants and Transition States from Methanol to Other Solvents[a]

Reaction	Solvent	$\Delta G_{tr}(RX)$	$\Delta G_{tr}(X^-)$	$\Delta G_{tr}(TS^{\ddagger})$
$CH_3Cl + N_3^-$	DMF	-0.5	6.7	1.6
$CH_3I + Cl^-$	DMF	-0.7	8.8	0.1
	H_2O	1.9	-3.4	-1.5
	CH_3CN	-0.5	8.6	1.8
$CH_3Br + SCN^-$	DMF	-0.4	3.7	1.0

[a]Data in kilocalories per mole obtained from solvent activity coefficients in A. J. Parker, *Chem. Rev.* **69**, 1 (1969).

other solvents. For example, it can be seen that transfer of the alkyl halides from methanol to DMF and acetonitrile is slightly exergonic, while transfer of the transition states from methanol to the dipolar aprotic solvents is slightly endergonic. However, the greatest contribution to the change in the free energy of activation, on transferring the reaction from methanol to a dipolar aprotic solvent, is manifested in the free energy of transfer of the anionic nucleophile to the dipolar aprotic solvent. This contribution is significantly endergonic (up to ~ 9 kcal mol^{-1}, which translates into rate changes as large as 10^6), so that weak solvation of the nucleophile in dipolar aprotic solvents is confirmed as being largely responsible for the large rate enhancements that are observed for these reactions in dipolar aprotic solvents. Note, by comparison, that when a nucleophilic substitution reaction is carried out in water, the free energy of transfer of reactants and transition state come out to be quite similar, so that the effect on the free energy barrier largely cancels out. Consequently, solvent effects in switching from one protic solvent to another turn out to be relatively small.

8.7 SOLVENT EFFECTS FROM THE CURVE-CROSSING MODEL

In Chapter 4 we showed how the barrier to a chemical reaction may be understood using the curve-crossing model. In this section we describe how solvent effects may be understood in curve-crossing terms—in particular, solvent effects on identity reactions, where simple thermodynamic arguments cannot be applied. Given that the energy barrier to a reaction can be dramatically increased (or decreased) by solvent, it should be possible to analyze the effect of solvent on barrier heights by exploring how solvent is likely to affect the various terms in the curve-crossing equation. Let us consider the identity S_N2 reaction of Eq. (8.1) to illustrate these ideas.

In qualitative terms what the solvent does is to add a *solvation barrier* on top of the intrinsic molecular barrier. There is a solvent barrier because it is not just the reactants $Cl^- + CH_3Cl$ that need to undergo structural rearrangement; the solvent needs to undergo reorganization as well. Initially the solvent is arranged around the reactants in such a configuration so as to solvate the reactants in the most effective way possible. This is particularly important around the charged chloride ion. However, as the reaction proceeds the initial solvation configuration is no longer satisfactory. The chloride anion is converted to a chlorine atom within a methyl chloride molecule so the initial strong solvation sphere around the chloride ion needs to be disrupted to accommodate the neutral methyl chloride molecule. In similar fashion the solvation sphere around the methyl chloride reactant needs to reorganize so it can solvate the chloride ion that is formed.

The energetic implications of this need to reorganize solvent can be assessed with the curve-crossing model. As we discussed in Chapter 4, the barrier to a chemical reaction $\Delta E^{\ddagger}$ is given by

$$\Delta E^{\ddagger} = fG - B \tag{8.17}$$

The barrier is just some fraction f of the initial energy gap G, between reactant and product configurations, which is reduced by the quantum mechanical mixing parameter B. For an S_N2 reaction in the gas phase, the magnitude of G is given by

$$G = I_N - A_{RX} \tag{8.18}$$

where I_N is the vertical ionization potential of the nucleophile, and A_{RX} is the vertical electron affinity of the substrate RX, both in the gas phase. If the reaction is carried out in solution, then the initial gap in solution $G(\text{sol})$ will just be

$$G(\text{sol}) = I_N^*(\text{sol}) - A_{RX}^*(\text{sol}) \tag{8.19}$$

where $I_N^*(\text{sol})$ and $A_{RX}^*(\text{sol})$ are now *the corresponding values in solution* (sol signifies a value in solution, and * signifies the term is a vertical one). These values have been estimated by Shaik et al.[14] in the gas phase and in solution (water and DMF) and some representative values appear in Table 8.7. Two points are immediately obvious. First, initial energy gaps in solution are much greater than in the gas phase, in many cases by about 100 kcal mol^{-1}. Given that typically one fourth of the initial gap contributes to the barrier height (the f factor), the above data suggest that barrier heights in solution could well be 20–30 kcal mol^{-1} *higher* than in the gas phase. This crude estimate is very close to Jorgensen's calculation in which the barrier for the chloride exchange reaction in aqueous solution was found to be 24 kcal mol^{-1} higher than in the gas phase.

A second point to note is that the difference between the energy gaps in H_2O and DMF, typically 20–30 kcal mol^{-1}, translates into a barrier difference of 5–8 kcal mol^{-1} (if we once again assume an f factor of one fourth). This

TABLE 8.7. Initial Gap Size ($I_X - A_{RX}$) in the Gas Phase and in Solution for Identity S_N2 Reactions: $X^- + CH_3X^a$

X	$I_X: - A_{RX}$ Gas Phase (kcal mol^{-1})	$I_N^*(sol) - A_{RX}^*(sol)$ H$_2$O (kcal mol^{-1})	DMF (kcal mol^{-1})
F	135	259	230
Cl	113	204	181
Br	99	179	160
I	81	150	137
CN	159	242	218
HO	109	225	196
HS	95	189	166

^{a}Data taken from S. S. Shaik, H. B. Schlegel, and S. Wolfe, *Theoretical Aspects of Physical Organic Chemistry. The S_N2 Mechanism*, Wiley-Interscience, New York, 1992, Table 5.16, p. 201.

range is very similar to the barrier differences observed for anionic S_N2 reactions in protic and dipolar aprotic solvents, discussed in Section 8.6. So it seems that the solvent effect on the rate of an anionic S_N2 reaction can be usefully accommodated within the framework of the curve-crossing model.

How can the different values of $I_X: - A_{RX}$ in the gas phase and in solution (Table 8.7) be understood? It turns out that the difference depends largely on the difference in the solution and gas-phase ionization potential of the nucleophile, $I_X^*(sol) - I_X$. This value can be estimated using a simple thermochemical cycle that relates $I_X^*(sol)$ to I_X.

$$X:^-(sol) \rightarrow X:^-(g) \qquad \Delta E = S_{X:} \qquad (8.20)$$

$$X:^-(g) \rightarrow X\cdot(g) + e^- \qquad \Delta E = I_{X:} \qquad (8.21)$$

$$X\cdot(g) \rightarrow X\cdot(sol) \qquad \Delta E = -S_{X\cdot} \qquad (8.22)$$

$$X\cdot(sol) \rightarrow X\cdot{}^*(sol) \qquad \Delta E = R_{sol} \qquad (8.23)$$

$$\overline{X:^-(sol) \rightarrow X\cdot{}^*(sol) + e^- \qquad \Delta E = I_X^*(sol) \qquad (8.24)}$$

Accordingly, we may write

$$I_X^*(sol) = I_{X:} + S_{X:} - S_{X\cdot} + R_{sol} \qquad (8.25)$$

where solvation terms S are defined as positive, and R_{sol} is the energy required to reorganize the solvent structure around $X\cdot$ from its equilibrium arrangement to its equilibrium arrangement about $X:^-$ (i.e., a higher energy *nonequilibrium* arrangement when it is around $X \cdot$). Since R_{sol} is itself a function of the solvating ability of the solvent, and $S_{X:}$ is small and may be ignored (solvation energies of an uncharged species are small), Eq. (8.25) may be simplified to

$$I^*_{X:}(\text{sol}) = I_{X:} + (1 + \rho)S_{X:} \qquad (8.26)$$

where ρ is a solvent parameter whose value normally lies between 0 and 0.56 (the value for water) and reflects how resistant the solvent is to restructuring. Polar solvents that are more highly structured have large ρ values, while non-polar solvents have very low ρ values (the value for benzene is 0.01).

What Eq. (8.26) indicates is that when an S_N2 reaction with an anionic nucleophile is carried out in solution, the contribution of the nucleophile to the initial energy gap *increases by at least the nucleophile solvation energy $S_{X:}$* (and for larger ρ values—up to 1.5 times the nucleophile solvation energy). So the curve-crossing treatment of the S_N2 reaction reaffirms the earlier conclusion that a major contribution to the barrier of an S_N2 reaction in solution is likely to be the solvation energy of the nucleophile. (For a discussion of solvent effects on nucleophilicity see Section 9.4.1.)

REFERENCES

1. D. J. McLennan, *Aust. J. Chem.* **31,** 1897 (1978). For a theoretical estimate that is almost identical see J. Chandrasekhar, S. F. Smith, and W. L. Jorgensen, *J. Am. Chem. Soc.* **106,** 3049 (1984).

2. B. D. Wladkowski and J. I. Brauman, *J. Phys. Chem.* **97,** 13158 (1993).

3. N. Menshutkin, *Z. Phys. Chem.* **5,** 589 (1890).

4. J. Chandrasekhar and W. L. Jorgensen, *J. Am. Chem. Soc.* **107,** 2974 (1985).

5. J. Hine and P. K. Mookerjee, *J. Org. Chem.* **40,** 292 (1975).

6. E. D. Hughes and C. K. Ingold, *J. Chem. Soc.* 244 (1935).

7. E. Grunwald and S. Winstein, *J. Am. Chem. Soc.* **70,** 846 (1948).

8. E. M. Kosower, *J. Am. Chem. Soc.* **80,** 3253 (1958).

9. K. Dimroth, C. Reichardt, T. Siepmann, and F. Bohlmann, *Ann. Chem.* **661,** 1 (1963); C. Reichardt, *Angew. Chem. Int. Ed. Engl.* **4,** 29 (1965).

10. T. W. Bentley and G. E. Carter, *J. Am. Chem. Soc.* **104,** 5741 (1982).

11. T. W. Bentley, F. L. Schadt, and P. v. R. Schleyer, *J. Am. Chem. Soc.* **94,** 992 (1972).

12. E. Grunwald, G. Baughman, and G. Kohnstam, *J. Am. Chem. Soc.* **82,** 5801 (1960).

13. A. J. Parker, *Chem. Rev.* **69,** 1 (1969).

14. For an analysis of solvent effects using the curve-crossing model, see S. S. Shaik, H. B. Schlegel, and S. Wolfe, *Theoretical Aspects of Physical Organic Chemistry. The S_N2 Mechanism*, Wiley-Interscience, New York, 1992, p. 191.

PART C

REACTION TYPES

9

NUCLEOPHILIC AND ELECTROPHILIC REACTIVITY

9.1 INTRODUCTION

The so-called *polar* reactions constitute one of the basic reaction classes in organic chemistry. Nucleophilic addition and substitution, electrophilic addition and substitution, hydride and proton transfer, and many elimination reactions, all form part of this important class. Given the emphasis that is placed on charge types when classifying these reactions, it is not surprising that the concepts of the *nucleophile* and the *electrophile* are central in organic mechanistic thinking.

Nucleophilic substitution reactions were first studied mechanistically about 60 years ago by Hughes and Ingold. They laid down the foundation to modern mechanistic organic chemistry with their kinetic classification of nucleophilic aliphatic substitution into two groups: S_N1 and S_N2. Over the years considerable effort has been invested in obtaining a more detailed mechanistic understanding of these reactions with much of the emphasis being placed on the precise nature of reactions in the S_N1–S_N2 borderline region. Remarkably, however, it was only in the last decade that an intimate relationship between S_N2 and electron-transfer mechanisms was uncovered. In fact the S_N2 mechanism is now commonly classified as a *type* of electron-transfer process, even though it is not an electron-transfer process in the conventional sense of that term. Similar developments have occurred in the field of electrophilic aromatic substitution. An intimate relationship between the conventional polar mechanism and electron transfer has been discovered in this area as well.

The importance of electron transfer in organic chemistry has changed dramatically over the last decade or two. Before that time the study of electron-transfer reactions focused on inorganic metal complexes, and it was in this area that much of the mechanistic framework was formulated. Within mainstream organic chemistry, electron-transfer reactions in that earlier period went almost unnoticed. However, this situation has changed dramatically and organic electron-transfer processes are now studied intensively. So in the space of less than a generation, electron transfer in organic chemistry has been transformed from being a mechanistic oddity to a ubiquitous process, whose essence is at the heart of many organic reactions.

In this chapter we present an overview of nucleophilic aliphatic substitution and electrophilic aromatic substitution, with particular emphasis on these more recent developments.

9.2   ELECTRON-TRANSFER NATURE OF THE S_N2 REACTION

In Section 4.4.1 we described the two most important VB configurations from which the S_N2 reaction profile (for an anionic nucleophile) may be constructed. These are Ψ_R, the reactant configuration

$$\Psi_R = N:^- \ R\cdot \ \cdot X \tag{9.1}$$

and Ψ_P, the product configuration

$$\Psi_P = N\cdot \ \cdot R \ :X^- \tag{9.2}$$

Inspection of these two configurations reveals (within the context of this simplified electronic description) that the nature of the electronic reorganization that takes place in the conversion of Ψ_R to Ψ_P is just the *shift of a single electron* from the nucleophile N to the leaving group X; the nucleophile has lost an electron while the leaving group has gained one. If we use the D—A (donor–acceptor) terminology to label Ψ_R as the DA configuration, then Ψ_P would be labeled D^+A^- (N^- is the donor D and R—X is the acceptor A).

Note that the term *electron shift* is used deliberately to distinguish it from the term *electron transfer*. An electron-shift reaction is one whose reactant and product configurations are related by a change in the position of a single electron (as is the case for Ψ_R and Ψ_P), yet *cannot* be classified as electron transfer. Electron-transfer reactions are defined as those in which *either* the oxidized form of the reductant (D^+), *or* the reduced form of the oxidant (A^-) (or both), are formed in the reaction. The simplest case, where both are formed, may be represented by

$$D + A \rightarrow D^+ + A^- \tag{9.3}$$

However, just *one* of the two species, D^+ or A^-, will be formed when the electron jump from D to A induces a spontaneous reaction in either D^+ or A^-. An example of this type of process is the reaction of an electron donor with an alkyl halide.

$$D + R-X \rightarrow D^+ + R^{\cdot} + X^- \qquad (9.4)$$

This reaction, which induces spontaneous fragmentation of the R—X species, is classified as an electron-transfer reaction even though $[R \dotdiv X]^-$ may not be formed; the fact that D^+ is formed in the reaction identifies the process as an electron-transfer reaction.

According to these definitions the S_N2 reaction

$$N^- + R-X \rightarrow N-R + X^- \qquad (9.5)$$

should not strictly be classified as an electron-transfer process, since neither $N^{\cdot}$ nor $[R \dotdiv X]^-$ are primary reaction products. Hence, we classify this type of reaction by the term *single electron shift*.

Often one finds the S_N2 reaction referred to as a *two-electron process*. This terminology stems from the curved arrow description of the S_N2 reaction

$$HO^- \quad CH_3-I \longrightarrow HO-CH_3 + I^- \qquad (9.6)$$

in which it appears that the nucleophile gives up *two* electrons to the substrate while the leaving group leaves with *two* electrons. However, this way of counting is just a mnemonic intended to keep track of the valence electrons; it does not provide a count of the number of electrons that are actually transferred. By assessing group charges in the species shown in Eq. (9.6), it is immediately apparent that the S_N2 process involves a formal switch of just a *single* electron— from the OH^- group to the I atom; the hydroxide ion has formally *lost* a negative charge (it has gone from a charge of -1 to 0), while the I atom has formally *gained* a negative charge (it has gone from a charge of 0 to -1). (For simplicity, the polar contributions to the C—I and C—O bonds are ignored.) So the temptation to think that *two* electrons (or four electrons?) have been rearranged, as is implied by the conventional curved arrow description, should be dismissed. The *curved arrow description is an electron bookkeeping device, and is not intended to represent a physical description of the electronic reorganization.*

What is it about an electron-*shift* process that makes it distinct from an electron-*transfer* process? As we already noted, in an electron-transfer process either the oxidant or the reductant (or both) are generated, at least as transient species. In an electron-shift process, however, neither reductant nor oxidant are formed; the relocation of the electron is accompanied by bond reorganization in both the donor and the acceptor. Old bonds are broken and new bonds are formed, and this overshadows the electron relocation aspect of the reaction. Inspection of the reactant and product configurations [Eqs. (9.1) and (9.2)] reveals that shifting the electron from the nucleophile to the leaving group has induced the R—X bond to *break* and the N—R bond to be *created*. In other words, the bond reorganization may be thought of as occurring *as a conse-*

quence of the electron shift. Thus it is the bond reorganization in both the donor and the acceptor that distinguishes the electron-shift process from a simple electron-transfer process. This explains why the single-electron-shift character of the S_N2 reaction is not immediately obvious—the breaking of an old bond and the making of a new bond *mask* the nature of the single electron shift that is at the root of the reaction.

There is a second important feature that distinguishes the electron-shift from the electron-transfer processes. Even though an electron-shift process, such as the S_N2 reaction, involves the shift of a single electron, *no radical species are formed at any stage of the reaction.* This is in direct contrast to electron-transfer reactions where odd electron species *must* be involved, either in the reactants or in the products.

At first sight it might seem strange that the shift of a single electron from a closed-shell nucleophile to a closed-shell substrate, does not create odd electron species at any stage. If, however, we look at the two configurations from which each point on the reaction surface is built up, this conclusion is understandable. Both reactant and product configurations, depicted in Eqs. (9.1) and (9.2), are *singlet* configurations; all the electrons in these two configurations are spin paired. *Therefore any point on the reaction surface, which is described by a linear combination of these two configurations, must also be a singlet configuration.*

The VB picture described above, which portrays an S_N2 reaction as a single electron shift process, is disarmingly simple. Unfortunately, this simplicity does detract somewhat from the accuracy of the physical picture. The implication that an electron shifts from the nucleophile to the leaving group, without involving the methyl group in between, as implied by Eqs. (9.1) and (9.2), is not entirely satisfactory. It comes out this way because the simplified VB picture we employ is based on just two configurations. A more realistic view of the electron-shift process may be obtained if we improve our simple VB picture with insights that are obtained from an MO perspective.

An MO description of the S_N2 reaction focuses on two key orbitals: an electron pair (the nucleophile HOMO) and the σ^*_{R-X} orbital (the substrate LUMO). In MO terms therefore, the electron shift from the nucleophile to the R—X substrate takes place from the nucleophile HOMO into the substrate LUMO. Thus, when using an MO description the reactant configuration (DA) is

$$\Psi_R = n^2\sigma^2 \tag{9.7}$$

and the product configuration (D^+A^-) is

$$\Psi_P = n^1\sigma^2\sigma^{*1} \tag{9.8}$$

where n is the nucleophile HOMO, σ is the σ_{R-X} bond, and the superscript represents the number of electrons in the orbital. If we represent the MO description of Ψ_P using a VB representation, we may write

$$\Psi_P = N \cdot (R \dotdiv X)^- \qquad (9.9)$$

where the odd electron is placed between R and X to signify a three-electron bond. In Chapter 3 we saw that this three-electron bonded structure may be described in terms of two VB hybrids. Thus,

$$(R \dotdiv X)^- = R \cdot \ :X^- \leftrightarrow R:^- \cdot X \qquad (9.10)$$

and therefore Ψ_P becomes

$$\Psi_P = N \cdot R \cdot \ :X^- \leftrightarrow N \cdot R:^- \cdot X \qquad (9.11)$$

The point is that the VB representation of the MO product configuration [Eq. (9.11)] is different from the simple VB representation [Eq. (9.2)]. The MO representation contains within it some mixing of the carbanion configuration $N \cdot R:^- \cdot X$. Since for regular alkyl halides the weight of $R:^- \cdot X$ in the three-electron bond description is small (X being considerably more electronegative than R; see Section 3.5), we can simplify this description by ignoring the second term. We are then left with our original VB product configuration, Eq. (9.2), but with a better appreciation of the fact that an additional VB configuration, the so-called carbanion configuration $N \cdot R:^- \cdot X$, has, for simplicity, been left out. Of course, cases where additional configurations, such as the carbanion configuration, do play an important role and cannot be ignored are common. Examples are discussed in Section 9.7.

A crucial question must now be asked. *Does the classification of the S$_N$2 reaction within a general class of electron-shift processes assist us in understanding S$_N$2 reactivity?* For only if the answer to this question is *yes*, is the exercise of modeling S$_N$2 in electron-transfer terms of any scientific merit. Recent study on the S$_N$2 mechanism has in fact revealed that an electron-transfer perspective on this well-known reaction mechanism does provide valuable new insights into S$_N$2 reactivity, its relationship to other reaction mechanisms, and into the question of nucleophilicity.

To describe these points, we first briefly need to survey an area that has traditionally been within the domain of inorganic reaction mechanisms—*inner-sphere and outer-sphere electron transfer*—since it has turned out that this mechanistic description is highly relevant to organic reactions as well.

9.2.1 Inner-Sphere and Outer-Sphere Electron Transfer

Electron-transfer (ET) reactions, depicted in general form by Eq. (9.3), have been found to proceed by two mechanistic pathways: *outer-sphere ET* and *inner-sphere ET*.[1] In the outer-sphere ET mechanism, depicted in Scheme 9.1, the donor and acceptor approach one another to form a precursor complex.

However, any overlap between donor and acceptor orbitals within that complex is weak. This weakness is illustrated in Scheme 9.1 by D and A being surrounded by their individual solvation spheres. At this stage electron transfer takes place to generate the D^+A^- complex, which then dissociates into free D^+ and A^- species. The main point here is that the donor and acceptor species undergo essentially no bonding interaction during the electron-transfer process. Within the context of inorganic metal complexes, where much of the electron-transfer mechanistic work was carried out, this means that the ligand spheres of the two complexes remain intact, hence the term outer sphere. An example of an outer-sphere ET reaction is

$$[Fe(CN)_6]^{4-} + [IrCl_6]^{2-} \rightarrow [Fe(CN)_6]^{3-} + [IrCl_6]^{3-} \qquad (9.12)$$

Scheme 9.1

where an electron has been transferred from an Fe^{II} complex to an Ir^{IV} complex to give Fe^{III} and Ir^{III} complexes. It is for outer-sphere ET that the Marcus theory of electron transfer was developed (described in Chapter 5), and central to the theory is the idea that the barrier to an outer-sphere ET process comes about through the distortion of the metal–ligand bonds and solvation shells.

The alternative mechanistic pathway for electron transfer is the inner-sphere mechanism, depicted in Scheme 9.2. In this mechanism, significant bonding interaction takes place between the donor and acceptor within the precursor complex, as depicted by the dashed lines between D and A within the DA complex. At this point electron transfer takes place to yield the D^+A^- complex, which then dissociates to give separate D^+ and A^- species. The distinguishing feature of this mechanism is *that the bonding interaction in the precursor complex stabilizes the electron-transfer transition state, and so speeds up the ET process.* This transition state stabilization contrasts with the outer-sphere pathway where a bonding interaction is absent. The inner-sphere ET mechanism is exemplified by the reaction

$$[(NH_3)_5CoCl]^{2+} + Cr^{2+}(aq) \rightarrow (NH_3)_5Co^{2+} + [ClCr(H_2O)_5]^{2+} \qquad (9.13)$$

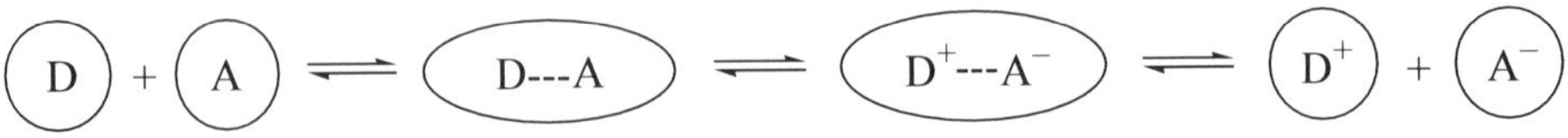

Scheme 9.2

This reaction is classified as inner-sphere because a bridged complex

$$[(NH_3)_5Co\text{---}Cl\text{---}Cr(H_2O)_5]^{4+}$$

9.1

is formed, in which the two metal ions, Co and Cr, share a common ligand chloride ion, and it is this bridging ligand that facilitates the electron transfer from Cr^{II} to Co^{III} (to give Cr^{III} and Co^{II}). In this case identifying the mechanism as inner sphere is easy—the chloride ligand is found to have switched from Co to Cr so the bridging complex **9.1** in which Cl is bonded to both metal ions is clearly implicated. However, a ligand transfer does not have to occur during an inner-sphere electron-transfer reaction; the inner-sphere mechanism may take place without leaving any tell-tale structural evidence.

A key feature that should always characterize the inner-sphere mechanism from the outer-sphere one is kinetic: the rate constant for the inner-sphere pathway is expected to be *greater* than that predicted by the Marcus equation, since Marcus theory was formulated to describe outer-sphere pathways. The Marcus treatment does not take into account the possibility of bonding interactions in the transition state, and since these stabilize the transition state, the inner-sphere reaction rate is increased in comparison to the Marcus prediction. Therefore this suggests *that the outer-sphere mechanism is only followed when an inner-sphere pathway—the one that would be energetically favored—is for some reason precluded.*

9.2.2 The S$_N$2 Reaction as an Inner-Sphere Electron Transfer

So what does this brief diversion into metal ion redox chemistry have to do with nucleophilic aliphatic substitution? Simply stated, if we apply the inorganic mechanistic terminology to the S$_N$2 reaction, we discover that *the S$_N$2 reaction may be characterized in a kinetic sense as a kind of inner-sphere electron-transfer process.* Thus the S$_N$2 reaction of hydroxide ion with methyl iodide, which proceeds through the transition state

$$[HO\text{---}CH_3\text{---}I]^-$$

9.2

may be thought of as an inner-sphere electron-transfer reaction from HO$^-$ to I that is facilitated by the bridging group CH$_3$. It is mechanistically akin to the reaction of Eq. (9.13), where an electron is transferred from Cr^{II} to Co^{III}, and which is facilitated by the bridging group Cl (as shown in **9.1**).

Of course the analogy in structure between **9.1** and **9.2** is not sufficient grounds for presuming a common mechanistic link between two seemingly very different reactions. Just as the relationship between inner-sphere and outer-sphere mechanisms has a *kinetic criterion* (viz., that the inner-sphere pathway is kinetically preferred over the outer-sphere pathway), so it is important to establish that the classification of the S_N2 reaction as a kind of inner-sphere ET reaction is useful; *that it has verifiable kinetic consequences.*

9.3

An elegant experiment that revealed this kinetic link and demonstrated that, mechanistically speaking, the S_N2 reaction *is* a type of electron-transfer process, was carried out by Lund and Lund.[2] These workers studied the rates of the nucleophilic reaction, shown in Eq. (9.14), for different alkyl bromides (R = 1-adamantyl, neopentyl, *tert*-butyl, *sec*-butyl, ethyl). The question posed by these workers was *How does the rate constant of an outer-sphere electron transfer from the enolate ion, **9.3**, to the substrate, R—Br, compare with the experimental rate constant for the actual nucleophilic substitution reaction between these two species, depicted in Eq. (9.14)? Is there any relationship between the two rate constants?*

This question can only be answered indirectly. It cannot be answered directly because an electron-transfer reaction from the enolate nucleophile to the alkyl bromides *simply does not occur*; the products of reaction of the enolate nucleophile with RBr substrates are substitution products and *not* electron-transfer products. Therefore, in order to answer this question, a way must be found *to estimate the rate constant for the hypothetical electron-transfer reaction.* In an ingenious experiment Lund and Lund *estimated* the rate of electron transfer from the enolate to each alkyl bromide electrochemically.[2] Let us describe their approach.

Lund and Lund first studied the rate of electron transfer from a family of aromatic radical ions to each of the alkyl bromides.

$$[\text{Aromatic}]^{\cdot-} + \text{R---Br} \xrightarrow{\text{ET}} \text{Aromatic} + \text{R} \cdot + \text{Br}^- \qquad (9.15)$$

Reduction, either chemical or electrochemical, of an alkyl halide is an example of a dissociative electron-transfer process. As discussed previously, species such as $[\text{R} \dot{-} \text{X}]^-$ are unstable so that the reduction of RBr leads to the formation of R $\cdot$ and Br$^-$, as shown in Eq. (9.15). Not surprisingly, the rate constant for electron transfer from the aromatic radical ion to the alkyl bromide is found to depend on the redox potential of the aromatic radical ion. An example of this type of relationship for a family of radical anions reacting with *tert*-butyl bro-

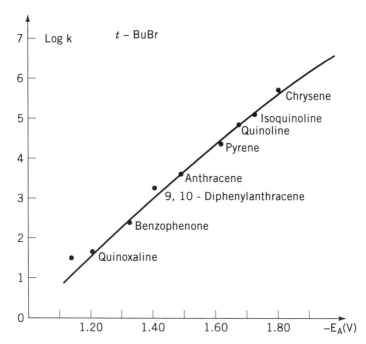

Fig. 9.1. Plot of the logarithm of the electron-transfer rate constant from a family of aromatic radical anions to *tert*-butyl bromide, versus the reduction potential E_A of the aromatic substrate. [Reproduced with permission from Fig. 2 in T. Lund and H. Lund, *Acta Chem. Scand., Ser. B*, **40**, 470 (1986). Copyright © Acta Chemica Scandinavica, 1986].

mide, as a function of the radical anion redox potential, is shown in Fig. 9.1. We see that the less stable the radical ion (i.e., the more negative the reduction potential), the faster the rate of electron transfer, just as one might expect.

Such an electron-transfer rate constant–driving force correlation for each alkyl bromide can now be used to interpolate (or extrapolate) the rate constant k_{ET} for an (hypothetical) electron-transfer reaction between *any* nucleophile and the particular RBr. The only additional information we need to carry out this interpolation is the redox potential of the nucleophile in question, which, for the nucleophile **9.3**, was found to be -1.13 V (vs Ag/AgI). Thus, for example, the extrapolated rate constant for electron transfer from **9.3** to *tert*-butyl bromide, as obtained from Fig. 9.1, is $12\ M^{-1}\ s^{-1}$. In this way the set of estimated k_{ET} values from **9.3** to the set of alkyl bromides was obtained. The results are listed in Table 9.1. Table 9.1 also lists the actual nucleophilic substitution rate constants k_{sub} obtained experimentally, together with the k_{sub}/k_{ET} rate ratios.

Inspection of the data in Table 9.1 is insightful. It shows that substitution and electron-transfer rate constants are indeed related. The data indicate that for 1-adamantyl bromide, neopentyl bromide and, to a lesser extent, *tert*-butyl bromide *the experimental substitution rate constants are very close to the cor-*

TABLE 9.1. Rate Constants for Nucleophilic Substitution k_{sub}, and Electron Transfer, k_{ET}, between 1,4-dihydro-4-methoxycarbonyl-1-methylpyridine, 9.3, and Alkyl Bromides, R—Br[a]

R—Br	k_{sub}	k_{ET}	k_{sub}/k_{ET}	ΔG_{stab}
1-Ad Br	1.5×10^{-2}	1.9×10^{-2}	0.80	≈ 0
Neopentyl Br	2.9×10^{-2}	2.3×10^{-2}	1.3	0.1
tert-Butyl Br	30	12	2.5	0.5
sec-Butyl Br	480	2.8	170	3.0
n-Butyl Br	1420	3.5	400	3.5
Ethyl Br	3052	1.2	2500	4.6

[a]Data taken from T. Lund and H. Lund. *Acta Chem. Scand., Ser. B*, **40**, 470 (1986).

responding estimated electron-transfer rate constants. Given that rate constants of chemical reactions can cover over 20 orders of magnitude, the similarity is unlikely to be coincidental. The similarity in rate constants strongly suggest that both processes share the same rate-determining step. In other words, *the mechanism of the substitution reaction for these three substrates is by rate-determining electron transfer.* The fact that the structure of 1-adamantyl, neopentyl, and *tert*-butyl groups precludes any significant nucleophile–alkyl group interaction means that the electron-transfer process is close to being outer sphere in nature; that is, there is very little, if any, nucleophile–carbon bond making in the electron-transfer transition state. The transition state for these reactions can therefore be depicted by **9.4**, where the outer-sphere character is illustrated by the discrete solvation spheres.

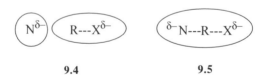

| 9.4 | 9.5 |

The remaining substrates, *sec*-butyl, *n*-butyl, and ethyl bromides, behave differently. We see from Table 9.1 that for these substrates, the nucleophilic substitution rate constant is substantially *larger* than the electron-transfer rate constant (the rate constant ratio k_{sub}/k_{ET} reaches 2500 for ethyl bromide). We already noted that nucleophilic substitution reactions are single-electron-shift processes, so the results suggest that the electron shift is in some way kinetically assisted compared to an outer-sphere electron transfer (such as from a radical anion). *Thus the nucleophilic substitution reactions of the unhindered alkyl bromides may be classified in a kinetic sense as inner-sphere electron-transfer reactions* (though as we mentioned earlier, these reactions should not be formally classified as electron transfer). The transition states for these reactions, depicted as inner-sphere ET, are illustrated by **9.5**. Thus the k_{sub} values are *larger* than the (hypothetical) k_{ET} values because in the S_N2 transition state, there is some nucleophile–carbon bonding that stabilizes the transition state and enhances the reaction rate. (Recall that the transition state of an S_N2 reaction can be conveniently expressed as a resonance hybrid of the two configurations Ψ_R and Ψ_P, that is,

$$\Psi_{TS} = 1/\sqrt{2}\{(N:^- \ R\cdot \ \cdot X) \leftrightarrow (N\cdot \ \cdot R :X^-)\} \qquad (9.16)$$

so that this nucleophile–carbon bonding comes about from the contribution of Ψ_P to the description of the transition state.)

The magnitude of this free energy stabilization ΔG_{stab}—a measure of the extent of nucleophile–carbon bonding in the transition state—can be readily estimated by substituting the k_{sub}/k_{ET} rate constant ratio into

$$\Delta G_{stab} = -RT \ln k_{sub}/k_{ET} \qquad (9.17)$$

where Eq. (9.17) is derived from the Eyring equation, Eq. (5.22). Values of ΔG_{stab}, calculated in this way, are shown in Table 9.1. In the hypothetical outer-sphere transition state, **9.4**, there is no stabilizing interaction between the electron donor and acceptor, so the rate of substitution and the rate of electron transfer would be the same!

Therefore, we learn from Lund and Lund's experiment that the rate of an S_N2 reaction (between N and RX) is related to the electron-transfer rate (from N to RX) in the following way:

$$k_{sub} \geq k_{ET}$$

The rate of the substitution reaction will be greater than or equal to the rate of the outer-sphere electron transfer.

9.6

The above discussion suggests, therefore, that there is an inner-sphere–outer-sphere mechanistic spectrum, depicted in **9.6**, which encompasses all electron-transfer processes, and S_N2 reactions lie at the inner-sphere end of that spectrum.[2] The precise position in this spectrum may be defined by the magnitude of the inner-sphere transition state stabilization ΔG_{stab}, and is determined by the intermolecular bonding interaction in the transition state. For the reactions shown in Eq. (9.14) the stabilization energy is always relatively small—even a rate enhancement of 2500 translates into a ΔG_{stab} of just 4.6 kcal mol^{-1} so that these reactions are either electron transfer or loose S_N2 with substantial ET character. In other cases, however, the k_{sub}/k_{ET} rate ratio can become very substantial. For example, applying the same treatment to the S_N2 reaction of ethyl bromide with iodide ion leads to a rate constant ratio of 4.10^{20}(!). This large rate enhancement translates into a ΔG_{stab} value of 29 kcal mol^{-1} suggesting very substantial transition state stabilization in this case.

Viewing the S_N2 reaction as an electron-shift process enables us to understand the S_N2 reaction kinetically. We will subsequently see that a single-electron perspective on this reaction provides additional mechanistic insights as well.

9.3 NUCLEOPHILICITY

A nucleophile is a Lewis base that uses an available electron pair to bond to carbon. The term *nucleophilic ability* is normally used in a kinetic sense; that is, a *good* nucleophile is one that forms a new bond to carbon *rapidly* (either with or without the expulsion of some leaving group). While the definition might appear to be straightforward, the question of what factors govern nucleophilicity has attracted considerable attention (and controversy) over the past half century. The only point on which full agreement has been reached is that the subject is complex, and that a well-defined and unique nucleophilicity scale is unachievable.

The reasons the problem of nucleophilicity is complex are easier to point out than to resolve. First, the nucleophilic order is *substrate* dependent. For example, the nucleophilic order of three common nucleophiles toward methyl iodide in aqueous solution is $CN^- > HO^- > N_3^-$, while toward a cationic electrophile (such as Ph_3C^+) the order is reversed, that is, $N_3^- > HO^- > CN^-$.[3] Second, the nucleophilic order is *solvent* dependent. For example, the halide nucleophilic order in protic solvents is $I^- > Br^- > Cl^-$, while in dipolar aprotic solvents the order becomes $Cl^- \approx Br^- > I^-$. These two facts alone indicate that any discussion concerning nucleophilicity must take into account both the substrate that is attacked, and the solvent in which the reaction is carried out; one cannot focus on the intrinsic properties of the nucleophile alone.

The first attempt to quantify nucleophilic reactivity was by Swain and Scott[4] who proposed that the nucleophilicity of a reagent n be defined within a linear free energy relationship

$$\log (k/k_0) = sn \tag{9.18}$$

where k is the rate constant for the S_N2 reaction of a substrate with a particular nucleophile at 25°C, k_0 is the corresponding rate constant with a standard nucleophile (chosen by Swain and Scott to be water), n is the nucleophilicity parameter for the particular nucleophile, and s is a sensitivity parameter (normally <1) that measures how sensitive the substrate is toward nucleophilic variation. The n values were measured using methyl bromide as a reference substrate (whose s value is defined as 1). Thus the Swain–Scott equation (for nucleophilicity) is analogous to the Hammett equation (for substituent effects) and the Winstein–Grunwald equation (for solvent effects) in their general form, and in their empirical approach to the parameter they attempt to quantify.

A revised nucleophilic scale based on methyl iodide as reference substrate ($s = 1$) and methanol as the reference nucleophile ($n = 0$) has since been proposed. A large number of n values are now available on this newer scale. A selection of n values is listed in Table 9.2. It should be remembered, however, that the figures in this table are only of quantitative value for S_N2 substrates; they do not apply to cation–nucleophile reactions or even to nucleophilic reactions on carbonyl derivatives.

What are the factors that govern nucleophilicity? Pearson[5] recently noted that as many as 17 factors have been identified as influencing nucleophilic reactivity. While this large number may provide a guide to the extent of interest in the subject and the creative ability of those working in the area, its generous proportions also accounts for the fact that we still cannot reliably predict the rate of reaction between a new nucleophile:electrophile pair. However, the situation is not all one of gloom; several key factors that appear to dominate nucleophilicity have been established.

One factor that was recognized many years ago is *basicity*. A plot of the rate constant for nucleophilic attack against the nucleophile pK_a for a set of nucleophiles often gives an excellent correlation. This is shown in Fig. 9.2 for

TABLE 9.2. Nucleophilicity Parameters n (Based on Reaction with Methyl Iodide in Methanol)^a

Nucleophile	n	Nucleophile	n
CH_3OH	0.0	$(CH_3)_2S$	5.54
F^-	2.7	Pyridine	5.23
Cl^-	4.37	N,N-Dimethyl-Cyclohexylamine	5.64
Br^-	5.79	Aniline	5.70
I^-	7.42	Thiophenol	5.70
CH_3O^-	6.29	NH_2OH	6.60
PhO^-	5.75	NH_2NH_2	6.61
$PhCOO^-$	4.5	$(Et)_3N$	6.66
CH_3COO^-	4.3	Ph_3P	7.00
CN^-	6.70	Piperidine	7.30
SCN^-	6.70	$(Et)_3P$	8.72
SO_3^{2-}	8.53	$(Et)_3As$	6.90
$S_2O_3^{2-}$	8.95		
PhS^-	9.92		
NO_2^-	5.35		
N_3^-	5.78		

^{a}Data taken from R. G. Pearson, H. Sobel, and J. Songstad. *J. Am. Chem. Soc.* **90**, 319 (1968).

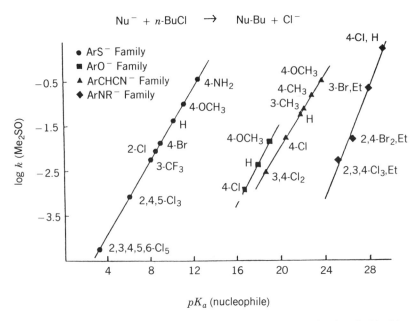

Fig. 9.2. Plot of rate constants for nucleophilic substitution of *n*-butyl chloride as a function of the nucleophile pK_a for different families of nucleophiles. [Reproduced with permission from Fig. 5 in F. G. Bordwell, T. A. Cripe, and D. L. Hughes, in *Nucleophilicity*, ACS Advances in Chemistry Series No. 215, J. M. Harris and S. P. Mc Manus, Eds. American Chemical Society, Washington, DC, 1987. Copyright © (1987) American Chemical Society.]

the reactions of *n*-butyl chloride. For example, there is an excellent correlation for a family of substituted thiophenoxides; the stronger base is the more powerful nucleophile. However, a linear correlation such as this only holds within a family of nucleophiles in which the donor atom remains the same. If one adds a family of substituted phenoxides to the plot, then again a linear correlation is obtained for this family, but on a line that is set apart from the thiophenoxide line and parallel to it.[6] In other words, nucleophiles that attack through different atoms do not obey a global nucleophilicity—basicity correlation; it is easy to find examples of weaker bases that are stronger nucleophiles (e.g., thiophenoxide is a stronger nucleophile than phenoxide by four orders of magnitude but a weaker base by some seven pK_a units). Clearly, basicity and nucleophilicity are governed not only by factors that affect both properties, but also by factors that are *different*. Given that nucleophilicity is a kinetic property while basicity is a thermodynamic property, the limited relation between the two should not be too surprising.

Let us now attempt to describe the factors that govern nucleophilicity with the aid of the curve-crossing model.

9.4 S_N2 REACTIVITY BASED ON THE CURVE-CROSSING MODEL

In order to understand the factors that govern nucleophilicity—indeed, to understand the factors that govern S_N2 reactivity as a whole—it is useful to build up an energy profile for the nucleophilic reaction, and examine the factors that determine the barrier height.

We have already pointed out that the barrier to an S_N2 reaction may be built up from the avoided crossing of reactant (**9.7**) and product (**9.8**) configurations (Sections 4.4 and 9.2). [Recall that **9.7** is termed the reactant configuration because it is the primary contributor to the electronic structure of the reactants (a lone pair on the nucleophile and a Heitler–London bond pair between R and X) and **9.8** is termed the product configuration since it is a good descriptor of the electronic arrangement in the products—a bond pair between N and R, and a lone pair on the leaving group X.] A curve-crossing diagram for the S_N2 reaction of RX (where R is a primary alkyl derivative) with some anionic nucleophile is illustrated in Fig. 9.3.[7] Additional configurations describing

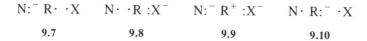

N:⁻ R· ·X	N· ·R :X⁻	N:⁻ R⁺ :X⁻	N· R:⁻ ·X
9.7	**9.8**	**9.9**	**9.10**

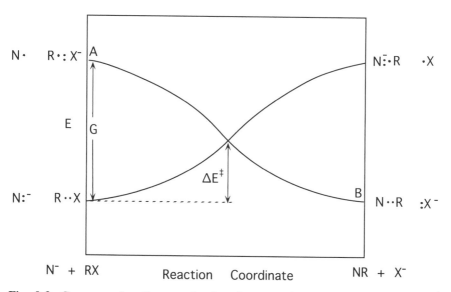

Fig. 9.3. Curve-crossing diagram showing the generation of an energy barrier $\Delta E^{\ddagger}$ for the S_N2 reaction: N:⁻ + R—X → N—R + X⁻ from the crossing of reactant N:⁻ R· ·X and product N· ·R :X⁻, configurations. (The effect of B, the quantum mechanical mixing parameter, is ignored for simplicity.)

potential intermediates—the carbocation configuration **9.9** and the carbanion configuration **9.10**—are ignored for the time being, since these are likely to be of less importance for primary alkyl derivatives, for which both R^+ and R^- are relatively unstable.

9.4.1 A Curve-Crossing Approach to Nucleophilicity

Inspection of Fig. 9.3 enables us to specify several factors that will affect nucleophilicity. What parameters associated with the nucleophile will determine the height of the energy barrier $\Delta E^{\ddagger}$? (For simplicity the barrier height is taken as the energy gap between reactants and the configuration-crossing point—the B term is ignored.) Nucleophilicity will be governed by a combination of (1) *kinetic* and (2) *thermodynamic* factors. Let us consider these in turn.

1. An important kinetic contribution to nucleophilicity is the initial energy gap. This is defined by the energy of the top-left-hand anchor point (labeled A in Fig. 9.3). Lowering this point from A_1 to A_2 (as indicated in Fig. 9.4a) will *decrease* the barrier height ($\Delta E_2^{\ddagger} < \Delta E_1^{\ddagger}$). Since the initial energy gap G for an S_N2 reaction in solution is determined by $I_{N:}^*(\text{sol}) - A_{RX}^*(\text{sol})$, the nucleophile contributes to the size of G by the magnitude of its solution vertical ionization potential [recall that $I_{N:}^*(\text{sol})$ is the solution vertical ionization potential of the nucleophile N: ; $A_{RX}^*(\text{sol})$ is the solution vertical electron affinity of the substrate RX; the symbol * signifies the energy term is vertical; and the label (sol) signifies the value is in solution.]

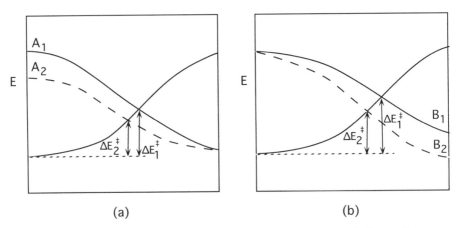

(a) (b)

Fig. 9.4. Curve-crossing diagram showing kinetic and thermodynamic factors that govern nucleophilicity. (a) Decreasing the nucleophile ionization energy and the nucleophile solvation energy decreases the initial energy gap (from A_1 to A_2) and is predicted to decrease the barrier height ($\Delta E_2^{\ddagger} < \Delta E_1^{\ddagger}$). (b) Changes in the nucleophile that increase reaction exothermicity (by lowering B_1 to B_2) are predicted to decrease the barrier height $\Delta E_2^{\ddagger} < \Delta E_1^{\ddagger}$.

Vertical solution ionization potentials $I_{N:}^*(sol)$ may be estimated through a thermodynamic cycle in which they are broken down into gas-phase ionization potentials and solvation energies. This was described in Section 8.7 and led to the following relationship:

$$I_{N:}^*(sol) = I_{N:} + S_{N:}(1 + \rho) \qquad (9.19)$$

where $S_{N:}$ is the solvation energy of the nucleophile (expressed as a positive number) and ρ is a solvent factor (< 1) that quantifies the "rigidity" of the solvent structure. By using this equation, vertical solution ionization potentials for anionic nucleophiles in H_2O and DMF have been estimated and a selection of values appears in Table 9.3 together with the corresponding gas-phase data. *Thus the model predicts that a nucleophile will be less reactive if, (a) it has a high gas-phase ionization potential and (b) if it is placed in a strongly solvating solvent.*

2. In addition to the kinetic affect on reactivity, there is also a *thermodynamic* effect that we need to consider. It is apparent that the nucleophile also affects the barrier height through the thermodynamic stability of the product (point B in Fig. 9.3). There are several factors that govern the position of point B on the energy scale. The most important of these are (a) the strength of the nucleophile–carbon bond that is formed; (b) the solvation energy of the nucleophile, which needs to be desolvated in order to form the N—R product

TABLE 9.3. Gas-Phase and Vertical Solution Ionization Potentials for Selected Anions[a]

Nucleophile	$I_{N:}$ Gas Phase	I_N^* (sol)	
		H_2O	DMF
F^-	78	240	213
Cl^-	83	204	183
Br^-	78	186	170
I^-	71	166	156
CN^-	89	203	181
HO^-	44	195	168
HS^-	54	177	156
N_3^-	62[b]	174[c]	155[c]

[a]Data taken from S. S. Shaik, H. B. Schlegel, and S. Wolfe, *Theoretical Aspects of Physical Organic Chemistry. The S_N2 Mechanism*, Wiley-Interscience, New York, 1992, Table 5.14, p. 198.
[b]Data from R. G. Pearson. *J. Am. Chem. Soc.* **108**, 6109 (1986).
[c]Calculated using Eq. (9.19), where the ρ value for H_2O is 0.56 and for DMF is 0.48. Solvation energies of N_3^- in H_2O and DMF are 72 and 63 kcal mol^{-1}, respectively.

(we are considering anionic nucleophiles here); and (c) the adiabatic ionization potential of the nucleophile, whereby the $N:^-$ electronic structure is converted to $N\cdot$, its product form. These three factors together affect the energy of point B (Fig. 9.3). If the energy of the product is lowered as illustrated in Fig. 9.4b ($B_2 < B_1$), then the barrier to reaction will also be reduced ($\Delta E_2^{\ddagger} < \Delta E_1^{\ddagger}$). This is just a further example of a rate–equilibrium relationship: The more stable product is expected to be formed more rapidly than the less stable product.

The above discussion allows us to conclude that the key *electronic* factors that govern nucleophilic reactivity are (a) the ionization potential of the nucleophile (adiabatic and vertical), (b) its solvation energy, and (c) the strength of the nucleophile–carbon bond. Of course, in addition to these electronic factors, we also need to consider any *steric* factors that will be superimposed on these electronic ones.

The reason that making predictions about nucleophilicity is difficult is now evident. Not only is nucleophilicity governed by a combination of several factors, but the *relative* contribution of each of these factors is substrate dependent. We would expect "early" transition states to be dominated by the *kinetic* factors (vertical nucleophile ionization potential and solvation energy) and less influenced by the solely *thermodynamic* factor: the nucleophile–carbon bond strength. On the other hand, "late" transition states would be more influenced by the thermodynamic factors and less by the kinetic factors. This change in the relative importance of the different factors means that the nucleophilic order is not expected to be invariant, but will be *substrate dependent*. Let us now consider some examples that illustrate these ideas.

1. *Why Is the Nucleophilic Order Substrate Dependent?* The nucleophilic order toward a cationic substrate, such as the pyronin cation **9.11**, for three common nucleophiles is $N_3^- > HO^- > CN^-$, while toward a regular S_N2 substrate such as methyl iodide, the order is reversed $CN^- > HO^- > N_3^-$.[3]

$$(CH_3)_2N \qquad O \qquad N(CH_3)_2$$
$$+$$

9.11

For these three nucleophiles the vertical solution ionization potential (from Table 9.3) decreases in the order $CN^- > HO^- > N_3^-$ (in both water and DMF). However, the nucleophile–carbon bond strength order decreases in the same order: CN^- (122 kcal mol^{-1}) $> HO^-$ (91 kcal mol^{-1}) $> N_3^-$ (70–80 kcal mol^{-1}). *Thus these two factors oppose one another.* The bond strength order favors CN^- as the best nucleophile, while the solution ionization potential order favors N_3^- as the best nucleophile (since high solution ionization potential leads to low nucleophilic reactivity). So the nucleophilic order will depend on the factor that is dominant. The observed nucleophilic order toward a cationic substrate, such as pyronin cation, appears to be governed by relative solution ionization potentials, while for a standard S_N2 substrate, such as methyl iodide, the reactivity order appears to be governed by the relative bond strengths.

Why do cations and neutral substrates respond differently? Nucleophilic attack on reactive carbocations is more exergonic than attack on typical S_N2 substrates (for cations there is no need to break a bond to a leaving group). Consequently, the transition state for attack on the reactive cations will be "early" and *will involve relatively little nucleophile–carbon bond making.* Reactivity in such systems is therefore dominated by the solution ionization potential, a kinetic factor. Indeed, one finds that the rates of nucleophilic attack of nucleophiles with cations correlates quite well with the nucleophile oxidation potential, which is just an electrochemical measure of solution ionization potential. For this reason the nucleophilic order toward cations is $N_3^- >$ HO$^-$ $>$ CN$^-$.

For S_N2 substrates, the reactions are less exergonic, the transition state is relatively "late" and therefore significant nucleophile–carbon bonding has taken place. (Recall that in our discussion of the S_N2 reaction as an inner-sphere electron transfer, it was estimated that the extent of transition state bonding for I$^-$ reacting with ethyl bromide was 29 kcal mol^{-1}.) Clearly, for S_N2 reactions with substantial nucleophile–carbon bond making in the transition state, the strength of the nucleophile–carbon bond will play an important role in governing reactivity. Thus the nucleophilic order toward S_N2 substrates (CN$^- >$ HO$^-$ $>$ N$_3^-$) follows the order of nucleophile–carbon bond strength.

2. *Why Is the Nucleophilic Order in Protic Solvents often Different from that in the Gas-Phase and Dipolar Aprotic Solvents?* The fact that the nucleophilic order is solvent dependent is well established. The nucleophilic order for halide ions in protic solvents [I$^- >$ Br$^- >$ Cl$^- >$ F$^-$ (see Table 9.2)] is quite different from the available data in dipolar aprotic solvents (Cl$^- \approx$ Br$^-$ $>$ I$^-$)[8] and from that in the gas phase (F$^- >$ Cl$^- >$ Br$^-$).[9] Let us analyze this problem in terms of reaction thermodynamics and initial energy gaps.

The nucleophile–carbon bond strength order is F$^- >$ Cl$^- >$ Br$^- >$ I$^-$ and the magnitudes vary substantially (C—F = 105 kcal mol^{-1}; C—Cl = 79 kcal mol^{-1}; C—Br = 67 kcal mol^{-1}; C—I = 57 kcal mol^{-1}) leading to major differences in the thermodynamic driving force. On the other hand, in the gas phase the ionization potentials of the halogens fall within the relatively narrow range of 71–83 kcal mol^{-1} (Table 7.5). As a result it is the bond strength differences that govern the reactivity order. *So the gas-phase nucleophilic order is just the bond strength order* (F$^- >$ Cl$^- >$ Br$^-$).

In solution, solvation effects on the gap size *and* on reaction thermodynamics need to be considered. As discussed above, the contribution of the nucleophile to the initial energy gap, $I_N^*(sol)$, incorporates the solvation energy of the nucleophile [Eq. (9.19)]. Since solvation energies for the halide ions differ significantly (see Table 8.1) and decrease in the order $F^- > Cl^- > Br^- > I^-$, initial energy gaps for the halides will also decrease in that order. The effect will be to favor the nucleophilic order $I^- > Br^- > Cl^- > F^-$. In solution, however, reaction thermodynamics also change compared to that in the gas phase due to differences in the solvation energies of the halide ions. Since solvation energies increase in the order $I^- < Br^- < Cl^- < F^-$ the effect of solvation on reaction thermodynamics will also favor the order $I^- > Br^- > Cl^- > F^-$.

The analysis suggests that both kinetic *and* thermodynamic solvation effects will favor the nucleophilic order $I^- > Br^- > Cl^- > F^-$, and oppose the intrinsic gas-phase order, which derives from the C—X bond strength. *If solvation effects are weak the bond strength order will be expected to dominate the nucleophilic order, while in cases where solvation effects are strong, solvation energy ordering will be expected to dominate the nucleophilic order.* This is precisely what is observed. In dipolar aprotic solvents where anionic solvation is relatively weak, the observed reactivity order $Cl^- \approx Br^- > I^-$ is close to the intrinsic gas-phase reactivity order, while in water, where anionic solvation is strong, the reactivity order $I^- > Br^- > Cl^- > F^-$ is governed by the solvation energy order and is the reverse of the gas-phase nucleophilic order.

9.5 RELATIONSHIP BETWEEN POLAR AND ELECTRON-TRANSFER REACTIONS

We noted that the importance of electron-transfer (ET) reactions in organic chemistry has grown enormously in recent years.[10] In fact, quite a few reactions, originally thought to proceed via a polar mechanism, are now thought to proceed via an electron-transfer mechanism.[10b] The reason for the ambiguity in some cases is now easier to see; energy profiles for both classes of reactions are constructed from the avoided crossing of DA and D^+A^- configurations, so both reactions share some common features, in particular their response to electronic effects. For this reason the two reactions are not always easy to distinguish from one another. Let us now consider the precise relationship between these two reaction types.

Consider the nucleophilic addition of hydroxide ion to a carbonyl group

$$HO^- \;+\; {\Large {}^{\backslash}_{\diagup}}C{=}O \;\longrightarrow\; HO{-}{\Large {}^{\backslash}_{\diagup}}C{-}O^- \tag{9.20}$$

Why does this reaction proceed via nucleophilic addition rather than by an electron-transfer process?[10a]

Reactant and product configurations for this reaction are shown in **9.12** and **9.13,** respectively. In the reactant configuration, labeled DA, the electrons are arranged so as to describe reactants—an electron pair on the hydroxide ion and

$$HO:^- \ \overset{\bullet}{C} - \overset{\bullet}{O} \qquad HO\bullet \ \bullet C - O:^-$$

$$DA \qquad\qquad D^+A^-$$

9.12   **9.13**

two odd electrons on C and O atoms that are coupled together in a carbonyl π bond. In the product configuration, labeled D^+A^-, an electron has been shifted from the hydroxide ion to the carbonyl group, so now there is an electron pair that bonds HO to C and a lone pair on the carbonyl oxygen. *The crucial point is that this configuration describes not only nucleophilic addition but electron transfer as well.* This is evident if we rewrite D^+A^- in a way that *separates* the HO group from the C—O group

$$HO\bullet \ \Big\} \ \bullet C - O:^-$$

9.14

Thus using DA terminology, the relationship between an electron-transfer reaction and a nucleophilic addition reaction may be described simply by

$$D + A \rightarrow D^+ + A^- \qquad \text{Electron transfer}$$

$$D + A \rightarrow D^+ - A^- \qquad \text{Nucleophilic addition}$$

In the electron-transfer reaction, the transfer of a single electron from D to A leads to D^+ and A^-. If D and A are even electron species then D^+ and A^- must be odd electron species. In the nucleophilic addition pathway, the electron shift *also leads to D^+ and A^- formation, but the two moieties are linked by a covalent bond.* Note that in the nucleophilic addition reaction, no odd electron species are formed at any stage (in contrast to the electron-transfer route); the wave function describing the reaction complex is a hybrid of **9.12** and **9.13**. Since both **9.12** and **9.13** are singlet wave functions (all electrons are spin paired), any linear combination of the two will also be a singlet wave function.

So the most direct way of distinguishing between these two reaction pathways is to establish if odd-electron species are formed during the reaction. The presence of odd-electron species is evidence for the electron-transfer route, while the absence of such species is suggestive of the polar pathway (though proving that something does *not* exist is harder than proving that it does!). However, a cautionary note is needed here; due to the high sensitivity of techniques for detecting odd-electron species (mainly EPR spectroscopy), the detection of odd-electron species in a system does not in itself *prove* an ET mechanism. It is very possible that the existence of small concentrations of odd-electron species may result from some minor unrelated reaction. Hence, for a proper mechanistic conclusion to be derived, it is essential to kinetically link the presence of the odd-electron species to the reaction in question.

Which of the two processes, nucleophilic addition or electron transfer, is likely to be preferred? The answer clearly depends on the identity of the particular donor and acceptor. However, some general guidelines may be laid down.

In enthalpic terms, nucleophilic addition is preferred over electron transfer. Because the nucleophilic addition process involves the formation of a new bond, the energy of the addition product D^+-A^- is likely to be more stable than the products of electron transfer D^+ and A^-. This is illustrated in Fig. 9.5 by the form of the two different D^+A^- curves. The nucleophilic addition curve comes down more steeply to give the more stable product so that the barrier for the addition process $\Delta E_{polar}^{\ddagger}$, is *smaller* than for the ET process $\Delta E_{ET}^{\ddagger}$. Entropic considerations, however, are likely to favor the ET pathway; whereas the nucleophilic addition process requires a relatively tight transition state in which incipient bond-making has taken place, an electron-transfer pathway involves a much looser transition state, in which overlap between D and A is much weaker.[10c]

The factors that are likely to encourage electron-transfer reactions over polar reactions can be described on the basis of the above analysis. In essence all revolve around the one basic principle: *Any factor (steric, electronic, or geometric) that will inhibit or hinder the bond coupling process will tend to encourage the entropically favored electron-transfer process over the enthalpically favored polar pathway.* Let us now discuss what these factors might be.

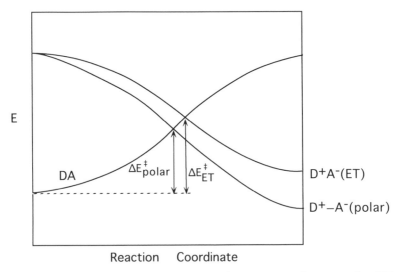

Fig. 9.5. Energy diagram showing the $DA-D^+A^-$ crossing for competing ET and polar pathways. $\Delta E_{ET}^{\ddagger} > \Delta E_{polar}^{\ddagger}$ due to the greater thermodynamic stability of the polar product.

9.5.1 Effect of Donor–Acceptor Ability

It is experimentally well established that the donor–acceptor pair ability has an important influence on the mechanistic route that is followed: ET or polar. When good donor–acceptor pairs react together, the ET process tends to predominate, while with poor donor–acceptor pairs the polar pathway tends to be preferred. The reaction of carbanions with carbocations illustrates the mechanistic dichotomy and is shown in Scheme 9.3.[11]

The carbanion R_1^- and the carbocation R_2^+ may combine directly to form the addition product R_1-R_2 (the polar route), or alternatively, electron transfer may take place between the carbanion and carbocation to give the radicals $R_1 \cdot$ and $R_2 \cdot$, which may then combine to give radical combination products R_1-R_1 and R_2-R_2, as well as the cross-products R_1-R_2. So the mechanistic route that is followed in any particular situation can be established from a product analysis.

The experimental observation is that relatively poor donor–acceptor pairs, such as the stable carbocation **9.15** and the carbanion **9.17** react via the polar pathway, while less stable carbocations, such as **9.16** (i.e., a better acceptor), react with the same carbanion via the ET pathway. A similar trend is noted for carbanions as well; the more stable carbanions prefer to react via the polar pathway, while the less stable ones tend to react via the ET pathway.

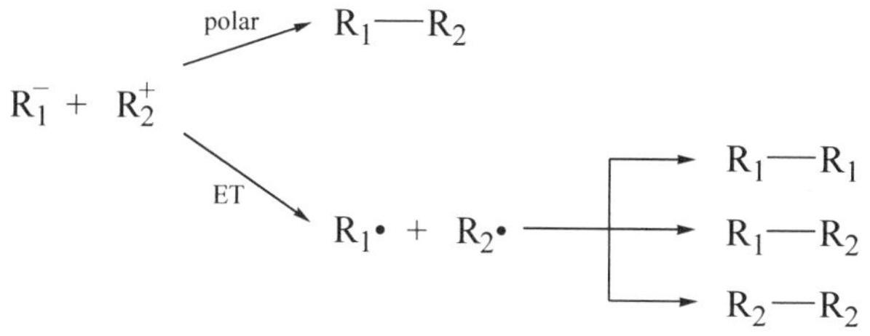

Scheme 9.3

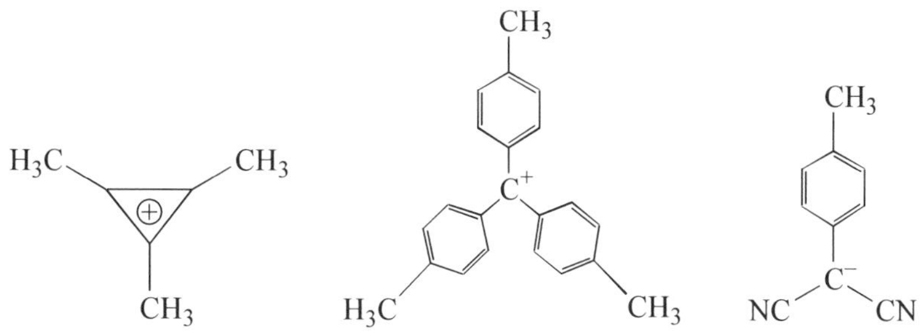

$E_{red} = -1.16$ $E_{red} = +0.05$ $E_{ox} = -0.33$

9.15   **9.16**   **9.17**

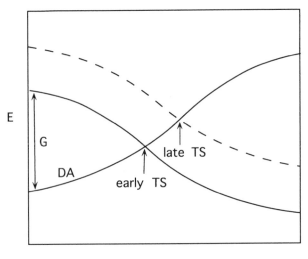

Reaction   Coordinate

Fig. 9.6. Curve-crossing diagram illustrating the effect of initial energy gap on the position of the transition state. Good donor–acceptor pairs (small G) will have earlier transition states than poor donor–acceptor pairs (large G).

The reason for this behavior can be understood with the aid of Fig. 9.6. The position of the transition state along the reaction coordinate depends on the donor–acceptor pair ability and reaction exothermicity; a good donor–acceptor pair exhibits a *small* initial energy gap G and large exothermicity, so the transition state is relatively *early*, while a poor donor–acceptor pair with a larger G and smaller exothermicity (whose D^+A^- curve is indicated by the broken line), exhibits a transition state that is relatively *late*. Now, an early transition state means that the extent of bond making is either small or non-existent. Accordingly, the entropically favored ET pathway predominates. On the other hand, a poor donor–acceptor pair with a relatively *late* transition state will have more substantial bonding between the donor and acceptor. This greater transition state stabilization, through a partial bond between the donor and acceptor, is likely to favor the polar pathway over the ET pathway.

9.5.2   Effect of Steric Interactions

The competition between polar and ET pathways is strongly influenced by steric factors. If either the donor or the acceptor moieties are sterically bulky then the ET pathway is likely to be preferred. For example, the nucleophilic addition of Grignard reagents to a bulky ketone, such as dimesityl ketone, appears to proceed via electron transfer [Eq. (9.21)], while the corresponding

$$RMgX + \quad \text{(anthraquinone/ketone structure with O)} \quad \xrightarrow{\text{ET}} \quad (RMgX)^{+\bullet} + \quad \text{(radical anion structure with } \overset{\bullet}{C}, O^-)$$

$$(9.21)$$

$$RMgX + CH_3COCH_3 \quad \xrightarrow{\text{nucleophilic addition}} \quad CH_3-\overset{\overset{\displaystyle R}{|}}{\underset{\underset{\displaystyle OMgX}{|}}{C}}-CH_3$$

$$(9.22)$$

reaction with unhindered ketones, such as acetone [Eq. (9.22)], involves direct addition of the Grignard reagent to the carbonyl compound.

Inspection of Fig. 9.5 makes it clear why steric factors favor the ET pathway. We already noted that the D^+A^- curve for the polar pathway comes down more steeply than that for the electron-transfer pathway due to partial bond formation between D and A in the transition state. For this partial bond formation to take place the two reacting species must approach one another to within bonding distance (normally <2 Å). If significant steric repulsions are present in the approach of D to A, then the energy lowering brought about by bond formation will be partially offset by the steric repulsions. The result is that the D^+A^- curve for the polar pathway will drop *less* steeply and the barrier to the polar mechanism will become *larger*. On the other hand, the electron-transfer pathway requires little overlap between donor and acceptor, so the reaction takes place at longer intermolecular distances. As a result the sensitivity of the electron-transfer mechanism to steric factors is small by comparison.

9.5.3 Effect of Donor–Acceptor Bond Strength

We have already noted that the factor that favors the polar over the ET pathway is the partial bond formation in the transition state. Accordingly, we would expect that the strength of the bond that is formed will have a significant bearing on whether the electron donor will react via a polar or ET mechanism. Indeed the reductant ability of the halogen ions (as reflected in their redox potentials in acetonitrile) increases in the order $F^- < Cl^- < Br^- < I^-$ *despite* the fact that nucleophilicity in aprotic solvents tends to increase in the reverse order. In other words, in dipolar aprotic solvents F^- is simultaneously the weakest reductant and the strongest nucleophile while I^- is the strongest reductant and the weakest nucleophile. The reason for the different polar–ET trends derives from the large change in the C—X bond strength that increases in the order

$C-I < C-Br < C-Cl < C-F$. Accordingly, polar processes are increasingly favored in that order due to the stabilizing effect of partial $C-X$ bond formation in the transition state.

The bond strength between two groups $D^{+\cdot}$ and $A^{-\cdot}$ and the degree of delocalization of the two radicals $D^{+\cdot}$ and $A^{-\cdot}$ are related. Bonds between *delocalized* radicals are significantly weaker than those between *localized* radicals. This is the reason the aliphatic $C-H$ bond strength in toluene (85 kcal mol^{-1}) is substantially weaker than that in methane (104 kcal mol^{-1}); the benzylic $C-H$ bond is formed by the coupling of the *delocalized* benzyl radical with a hydrogen atom while methane is formed through coupling of a *localized* methyl radical with hydrogen. Accordingly we would expect that electron transfer between D and A will be favored when the resultant radicals $D^{+\cdot}$ and $A^{-\cdot}$ are *delocalized*. Indeed, radical delocalization is one of the main factors that facilitates the well-established $S_{RN}1$ mechanism of nucleophilic substitution. Let us consider this mechanistic pathway in some detail.

9.6 $S_{RN}1$ MECHANISM

The $S_{RN}1$ mechanism for nucleophilic substitution, like the S_N1 and S_N2 reactions, involves the displacement of a leaving group by a nucleophile

$$R-X + :Y^- \rightarrow R-Y + :X^- \tag{9.23}$$

However, this reaction proceeds via a radical chain mechanism that is initiated by an electron-transfer step.[12] The important steps of the chain are depicted in Eqs. (9.24–9.27).

Initiation Step

$$R-X + e^- \rightarrow (R-X)^{\cdot-} \tag{9.24}$$

Propagation Steps

$$(R-X)^{\cdot-} \rightarrow R\cdot + :X^- \tag{9.25}$$

$$R\cdot + :Y^- \rightarrow (R-Y)^{\cdot-} \tag{9.26}$$

$$(R-Y)^{\cdot-} + R-X \rightarrow R-Y + (R-X)^{\cdot-} \tag{9.27}$$

If the nucleophile (Y^-) is a powerful reducing agent, the initiation step [Eq. (9.24)] may occur by electron transfer from Y^- to $R-X$. However, more typically the initiation step is carried out from some "external" source—either photochemically, electrochemically, or through the addition of some good electron donor in catalytic amount. An example of a reaction that proceeds via the $S_{RN}1$ mechanistic pathway is

$$\underset{NO_2}{\overset{H_3C}{\underset{}{H_3C}}}\!\!-\!\!C^- + CH_2\!-\!Cl \quad\overset{}{\longrightarrow}\quad \underset{NO_2}{\overset{H_3C}{\underset{}{H_3C}}}\!\!-\!\!C\!-\!CH_2 + Cl^-$$

(9.28)

What is it about this reaction that leads to a different mechanistic pathway? The answer lies in the stabilizing effect of the NO_2 group on the energies of the reactant and product radical ions $(R-X)^{\cdot-}$ and $(R-Y)^{\cdot-}$ that are generated along the S$_{RN}$1 pathway. In particular, the stability of $(R-Y)^{\cdot-}$ appears to be especially important since it is this stability that is responsible for a large driving force, and hence a large reaction rate, for the coupling reaction of $R\cdot$ with Y^-.[12c] The S$_{RN}$1 mechanism will only compete effectively with the S$_N$2 mechanism when the composite rate constant for the processes shown in Eqs. (9.25–9.27) will be sufficiently large. Let us discuss the nature of the radical anions $(R-X)^{\cdot-}$ and $(R-Y)^{\cdot-}$ in more detail.

We stated previously that an S$_N$2 reaction may be thought of as taking place by an electron shift from the nucleophile HOMO into the substrate LUMO: the σ^*_{R-X} orbital. However, substrates that undergo an S$_{RN}$1 substitution reaction are generally characterized by the fact that the LUMO is not the σ^*_{R-X} orbital. Instead it is usually a low-lying π^* type orbital. In the example of Eq. (9.28) the π^* orbital is delocalized over the nitrophenyl moiety and is unusually low in energy due to the strong electron–withdrawing ability of the NO_2 group (this accounts for the high electron affinity of many nitro substituted substrates). It is this low-lying orbital that results in the relative stabilities of $(R-X)^{\cdot-}$ and $(R-Y)^{\cdot-}$ species and enables the S$_{RN}$1 mechanism to compete effectively with the concerted S$_N$2 pathway.

The reason the initial electron transfer [Eq. (9.24)] and the first step in the propagation sequence [Eq. (9.25)] for S$_{RN}$1 substrates occur in two separate steps, rather than in a single step [as depicted in Eq. (9.29)], can be understood with the aid of a curve-crossing diagram, illustrated in Fig. 9.7.

$$R-X + e^- \rightarrow R\cdot + :X^-$$

(9.29)

The diagram involves three energy curves: the $R-X$ curve; the $(R-X)^{\cdot-}(\pi^*)$ curve for the radical anion state in which the added electron is in the π^* orbital; and $(R-X)^{\cdot-}(\sigma^*)$, in which the added electron is in the σ^* orbital. Note that in the reactant geometry (where no $R-X$ dissociation has taken place), $(R-X)^{\cdot-}(\pi^*)$ is lower in energy thatn $(R-X)^{\cdot-}(\sigma^*)$ (since π^* is lower in energy than σ^*). However, as the $C-X$ bond is elongated the π^* and σ^* curves cross over. This occurs because $R-X$ extension of the σ^* state

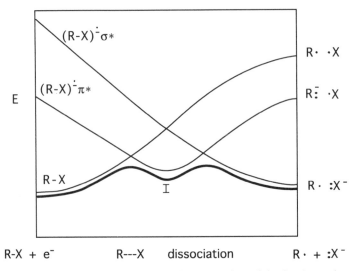

Fig. 9.7. Curve-crossing diagram showing the generation of the $S_{RN}1$ reaction pathway for nucleophilic substitution, via a radical anion species I (see text).

leads to the formation of the *more* stable pair $R\cdot$ and $X:^-$, while $R—X$ extension of the π^* state would lead to the *less* stable pair $R:^-$ and $\cdot X$. In the absence of the π^* orbital there would be an avoided crossing between the $R—X$ and $(R—X)^{\cdot-}\sigma^*$ curves, resulting in a concerted pathway for $R—X$ dissociation. However, the presence of the low-lying $(R—X)^{\cdot-}\pi^*$ curve leads to the formation of a lower energy *two-step* pathway for $R—X$ cleavage. The barrier for the first step (Fig. 9.7) is generated from the avoided crossing of the $R—X$ dissociation curve with the $(R—X)^{\cdot-}\pi^*$ curve, leading to the intermediate formation of a π^* radical anion $(R—X)^{\cdot-}$, labeled I. On proceeding along the reaction coordinate (by the additional stretching of the $C—X$ bond), a second avoided crossing between $(R—X)^{\cdot-}\pi^*$ and $(R—X)^{\cdot-}\sigma^*$ takes place, that is, an electron shifts from the π^* orbital into the σ^* orbital. This shift finally leads to $C—X$ bond fission.

The idea of two unoccupied orbitals competing to trap an added electron can be useful in analyzing the relationship between competing reaction mechanisms. It turns out the nature of the orbital into which the electron relocates may well determine the reaction mechanism that is followed. For example, consider the mechanisms for nucleophilic vinylic substitution. Vinyl systems commonly undergo nucleophilic substitution [Eq. (9.30)] via an addition–elimination pathway, rather than by the more direct S_N2 pathway.[13a] Why is this so?

$$\ce{>C=C<^{X} + Y^- -> >C=C<^{Y} + X^-} \tag{9.30}$$

An S_N2 process would require a single-electron shift into the σ^*_{C-X} orbital (as described previously for $R-X$) while the addition step requires the electron to shift into the π^* orbital. So the orbital that "wins" the electron will determine the reaction mechanism that is followed. If activating groups that lower the π^* orbital are present in the alkene (i.e., electron-withdrawing groups in the β positions), then attack occurs on π^* and the addition–elimination mechanism is preferred. If, however, no activating groups in the β position are present, and a leaving group with a low-lying σ^* orbital is employed, then normal back-side S_N2 substitution may take place.

Recent work has supported this view.[13b–d] When the leaving group in the vinyl system was $-IPh^+$, a group with a particularly low-lying σ^* orbital ($C-I$ bonds have low-lying σ^* orbitals; having a positively charged I atom would lower the σ^* orbital even further), nucleophilic attack by halide ion led to 100% *inversion* of configuration, suggesting that in this case substitution has proceeded by an S_N2 mechanism [Eq. (9.31)].[13b]

$$
\begin{array}{c}
\underset{n\text{-}C_8H_{17}}{\overset{H}{\diagdown}}C{=}C\underset{H}{\overset{I^+C_6H_5}{\diagup}} + X^- \longrightarrow \underset{n\text{-}C_8H_{17}}{\overset{H}{\diagdown}}C{=}C\underset{X}{\overset{H}{\diagup}} + C_6H_5I
\end{array}
$$

$$(9.31)$$

Similarly, it has recently been found that 1,2-dibromo-1,2-difluoroethylenes appear to undergo nucleophilic substitution with *complete* inversion of configuration, suggesting that an S_N2 pathway is followed in this case as well.[13c,d] Thus, it now appears than an S_N2 reaction at vinyl centers is more facile than previously thought.

9.7 S_N2 MECHANISTIC SPECTRUM

The above discussion has demonstrated the existence of an S_N2–ET mechanistic link and described the factors that are likely to favor one mechanism over the other. However, the S_N2 mechanistic pathway may also be extended in different directions—toward the two-step S_N1 pathway, and toward a carbanionic S_N2 process. Let us consider these mechanistic variations and see how they come about.

We previously described the S_N2 reaction in terms of an avoided crossing between two configurations, reactant **9.7** and product **9.8** (p. 235). However, two additional configurations, the carbocation configuration **9.9** and the carbanion configuration **9.10**, which describe potential intermediates, may also play a role in those cases where either of these intermediates are relatively stabilized. For S_N2 reactions on simple alkyl derivatives (e.g., CH_3-X), the role of **9.9** and **9.10** was ignored; CH_3^+ and CH_3^- are unstable species. However, these intermediate configurations may play an important role in systems with substituents on R that stabilize either R^+ or R^-.

Consider the substitution reaction on methoxymethyl derivatives

$$N:^- \ + \ \underset{\substack{| \\ OCH_3}}{CH_2} {-} X \quad\longrightarrow\quad \underset{\substack{| \\ OCH_3}}{N{-}CH_2} \ + \ X:^- \qquad (9.32)$$

In this case the carbocation is strongly stabilized due to the resonance inter-action

$$CH_3O{-}\overset{+}{C}H_2 \ \longleftrightarrow \ CH_3\overset{+}{O}{=}CH_2$$

However, the reaction remains S_N2 in character (the rate is nucleophile dependent).[14]

For such a reaction three (and not two) configurations are needed to properly describe the ground-state reaction surface. Hence, the transition state may be depicted in terms of *three* resonance structures

$$TS \ \equiv \ N:^- \ R \cdot \ \cdot X \ \leftrightarrow \ N \cdot \ \cdot R :X^- \ \leftrightarrow \ N:^- \ R^+ :X^- \qquad (9.33)$$

that is, the transition state is expected to exhibit some carbocationic character. Is there any evidence for the mixing-in of the carbocationic configuration into the TS wave function? A number of reactivity features can be explained on the basis of the above description.

First, according to the curve-crossing model, the mixing-in of some intermediate configuration will both stabilize the transition state (i.e., increase reactivity) as well as impart the character of that intermediate to the transition state. A major difference between the reaction of methyl derivatives and methoxy-methyl derivatives is that the latter are considerably more reactive, as would be expected from the above description. Second, when the Brønsted parameter, β_{nuc} for this reaction is measured (from a Brønsted plot of log k_{nuc} vs. pK_{nuc}) a particularly low value of 0.14 is found. In other words the rate is only *weakly* dependent on the substituent in the nucleophile. Also, when β_{lg} (lg $\equiv$ leaving group) is measured (from a plot of log k_{lg} vs pK_{lg}) large absolute values of between -0.7 and -0.9 are measured; in this case a *large* dependence of substituent on the leaving group ability is found.[14]

If we examine the resonance mix in the transition state [Eq. (9.33)], we see that mixing in the carbocationic contribution makes the transition state charge on the nucleophile *more* reactant-like (two out of three of the structures have the nucleophile in *reactant* form, $N:^-$) so a Brønsted plot shows a *smaller* dependence on the nucleophile. Another way of putting this is to say that since the transition state is reactant-like in terms of the nucleophile, substituent effects on the nucleophile are relatively small. The same argument suggests that the transition state charge on the leaving group is expected to be *large* (two out of three of the resonance contributors have the leaving group in its *product* form, $X:^-$) so substituent effects on the leaving group are relatively large and this is manifested in a *large* Brønsted slope. We learn from this analysis that when a third configuration contributes strongly to a description of the transition state, the Brønsted coefficients can provide a relative measure of charge distribution in the transition state, compared to the reference system (in this case CH$_3$X). Note, however, that the Brønsted coefficients should *not* be interpreted as a literal measure of TS charge—the factors that govern the actual magnitude of β are complex and include curve slopes, the quantum mechanical mixing parameter *B*, and not just the contribution of each configuration to the state description.

Just as electron-donating substituents on R stabilize the carbocation configuration, the introduction of electron-withdrawing substituents on R is expected to stabilize the carbanion configuration. The reactivity properties of alkyl derivatives activated by an α-carbonyl substituent

$$
\begin{array}{ccc}
\text{N:}^- \;+\; \underset{\substack{|\\ \text{RC}=\text{O}}}{\text{CH}_2}\text{—X} & \longrightarrow & \underset{\substack{|\\ \text{RC}=\text{O}}}{\text{N—CH}_2} \;+\; \text{X:}^-
\end{array}
\qquad (9.34)
$$

may be described by the configuration mix

$$
\text{TS} \equiv \text{N:}^- \;\text{R}\cdot\;\cdot\text{X} \leftrightarrow \text{N}\cdot\;\cdot\text{R}\;\text{:X}^- \leftrightarrow \text{N}\cdot\;\text{R:}^-\;\cdot\text{X} \qquad (9.35)
$$

since the carbanion configuration is stabilized through enolate delocalization.

$$
\underset{\substack{||\\ \text{O}}}{^-\text{C—C}} \longleftrightarrow \underset{\substack{|\\ \text{O}^-}}{\text{C}=\text{C}}
$$

9.18

Stabilization of the carbanion configuration is therefore expected to lead to an increase in reactivity and this has been confirmed. In fact, the relative reactivity of a series of RCH$_2$Cl with KI in acetone (S$_N$2 conditions) follows the same sequence as the acidity of RCH$_3$, strongly suggesting that the carbanion configuration plays a dominant role in governing reactivity.[15] The relevant data are listed in Table 9.4; substituents that increase the stability of RCH$_2^-$, as reflected in either low RCH$_3$ pK_a or high $\delta\Delta G°$ values, increase the S$_N$2 reactivity of RCH$_2$Cl.

Ab initio calculations on the S$_N$2 transition state confirm the importance of enolate resonance. The energy of the S$_N$2 transition state has been calculated with an α-CHO substituent in two conformations: the ''on'' conformation (**9.19**), where the resonance delocalization of the negative charge associated with the N---C---X system into the carbonyl group is possible; and the ''off'' conformation (**9.20**), where overlap of the carbanionic charge into the carbonyl group is not possible. The calculations show that the ''on'' conformation is significantly more stable than the ''off'' conformation.[16] Moreover, in this more

TABLE 9.4. Relative Reactivity of RCH$_2$Cl with KI in Acetone, Solution Acidities of RCH$_3$, and Relative Gas-Phase Acidities of RCH$_3$^{a}

R	Relative Substitution Rate with KI	pK_a of RCH$_3$ (DMSO)	$\delta\Delta G°$ (acid) (kcal mol^{-1})
Et—	1		
Ph—	250	40.9	36.3
EtOCO—	1,600		44.3
NC—	2,800	31.3	44.2
CH$_3$CO—	33,000	26.5	47.0
PhCO—	97,000	24.7	52.2

^{a}Data taken from D. J. McLennan and A. Pross, *J. Chem. Soc. Perkin Trans.* 2, 981 (1984).

stable conformation, the C—C bond is found to be shorter than a regular C—C bond, while the C—O double bond is slightly lengthened, once again reaffirming the importance of enolate resonance (**9.18**).

Delocalization "on"   Delocalization "off"

9.19   **9.20**

Some unusual characteristics of benzyl systems can now be understood. Benzyl systems are well-known for the fact that a plot of rate constant for nucleophilic substitution as a function of the substituent σ value for ring substituents, leads to a curved plot, often U shaped in form (Fig. 9.8).[17] For example, the nucleophilic substitution of substituted benzyl derivatives by aryl thiolate nucleophiles

$$\text{ArS}^- + \text{XC}_6\text{H}_4\text{CH}_2\text{—Br} \longrightarrow \text{XC}_6\text{H}_4\text{CH}_2\text{—SAr} + \text{Br}^- \quad (9.36)$$

conforms to this kind of pattern; both electron-releasing (e.g., OCH$_3$) and electron-withdrawing (e.g., NO$_2$) substituents in the ring *increase* reactivity compared to the unsubstituted parent. This contrasts with the normal pattern of substituent effects on reactivity, where electron-releasing and electron-withdrawing substituents have *opposite* effects on reactivity. What is the reason for this unusual behavior?

The benzyl system is unusual in that both the carbocation configuration **9.9** and the carbanion configuration **9.10** may play an important role in describing the reaction surface; if, for example, a *p*-NO$_2$ substituent is introduced into the ring then the carbanion configuration is stabilized, while if *p*-OCH$_3$ is intro-

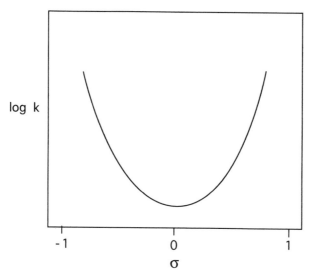

Fig. 9.8. Schematic representation of curved Hammett plot for the nucleophilic substitution reaction of ring-substituted benzyl derivatives. Both electron-withdrawing and electron-releasing substituents in the ring increase the rate of substitution compared to the unsubstituted parent.

duced into the ring then the carbocation configuration is stabilized. Thus for both electron-withdrawing *and* electron-releasing groups an ''intermediate'' configuration is stabilized, which leads to a rate enhancement compared to the unsubstituted parent, and hence, a curved Hammett plot.

The effect of substituents in the nucleophile are consistent with this picture. When a family of substituted anilines are reacted with the different benzyl derivatives

$$XC_6H_4-NH_2 + p\text{-}CH_3O-C_6H_4CH_2-Cl \longrightarrow$$

$$p-CH_3O-C_6H_4CH_2-NH-C_6H_4X + HCl \qquad (9.37)$$

$\rho = -0.59$

$$XC_6H_4-NH_2 + p\text{-}O_2N-C_6H_4CH_2-Cl \longrightarrow$$

$$p-O_2N-C_6H_4CH_2-NH-C_6H_4X + HCl \qquad (9.38)$$

$\rho = -1.55$

linear Hammett plots are obtained. However, ρ values for the different benzyl derivatives vary considerably, ranging from $\rho = -0.59$ for the p-OCH$_3$ derivative to $\rho = -1.55$ for the p-NO$_2$ derivative. For the p-NO$_2$ substituent the transition state is described by a resonance mix that includes carbanionic character [Eq. (9.35)], since the carbanion configuration is stabilized by the electron-withdrawing nitro group. Given that the carbanion configuration incorporates the nucleophile in its product form (i.e., $N \cdot$), an *increase* in the absolute magnitude of the ρ value is expected, as observed. For the p-OCH_3 substituent, it is the carbocation configuration that is stabilized and contributes significantly to the transition state wave function [described by Eq. (9.33)]. Since this configuration describes the nucleophile in its reactant form (i.e., N:), a *decrease* in the absolute magnitude of the ρ value is observed. For the unsubstituted benzyl derivative, where neither carbanion nor carbocation are expected to be especially important, an intermediate value of ρ is found.

9.8 ELECTROPHILIC REACTIVITY

Electrophilic aromatic substitution is one of the fundamental reactions of organic chemistry, yet its mechanistic description, like that of nucleophilic aliphatic substitution, is currently undergoing reassessment.

The traditional mechanism for electrophilic aromatic substitution may be represented by the two-step scheme

$$\text{Ar}-\text{H} + \text{E}^+ \rightleftharpoons \text{Ar}\overset{+}{\underset{}{\diagdown}}\begin{smallmatrix}\nearrow\text{H}\\\searrow\text{E}\end{smallmatrix} \longrightarrow \text{Ar}-\text{E} + \text{H}^+ \quad (9.39)$$

where ArH is the aromatic substrate and E^+ is a general electrophile. In the first step the electrophile attacks the aromatic substrate to form a σ complex (the so-called Wheland intermediate) and in the second step the intermediate eliminates H^+ to yield the substituted product. However, a number of features of this reaction were recently discovered, which suggested that this simple two-step polar mechanism requires refinement, and in some cases may actually need to be modified. For example, the mixing of an arene with an electrophile often appears to generate two complexes (not just one) between the reagents. Thus, in addition to the σ complex mentioned above, evidence for π-complex formation between the electrophile and the aromatic substrate was also obtained. Does the π complex lie on the electrophilic substitution reaction pathway? Or is the π complex merely a transient side product that forms reversibly during the reaction.

A recent important development in the chemistry of electrophilic aromatic substitution is the proposal that the thermal reaction leading to formation of the Wheland intermediate may in certain cases be initiated by an electron-transfer step

$$\text{Ar}-\text{H} + \text{E}^+ \xrightarrow{\text{ET}} \text{Ar}-\text{H}^{+\bullet} + \text{E}^\bullet \longrightarrow \text{Ar}\overset{+}{\underset{}{\diagdown}}\begin{smallmatrix}\nearrow\text{H}\\\searrow\text{E}\end{smallmatrix} \quad (9.40)$$

Thus the Wheland intermediate is postulated as being formed in a two-step, rather than a one-step, pathway. This view has been advocated by Kochi[18] following his observation that aromatic nitration can be brought about photochemically through the excitation of a charge-transfer π complex of an aromatic substrate and a nitrating agent NO_2Y (for a variety of different Y groups), as depicted in Scheme 9.4. The key feature of this work was Kochi's claim that for a number of systems studied, the isomer distribution from the photochemical pathway was the same as that observed in the conventional thermal nitration reaction. If photochemical excitation initiates the reaction by inducing electron transfer between the aromatic and the nitrating agent [Eq. (9.42)], then the similar product distribution for the thermal and photochemical pathways might suggest *that the thermal pathway also proceeds via an initial electron-transfer step*. While mechanistic details for some of these systems remain the subject of current debate (see below), what is already clear is that an electron-transfer perspective on this reaction provides a useful framework for considering this general mechanistic class.

$$\text{ArH} + \text{NO}_2\text{Y} \; \underset{}{\overset{K}{\rightleftharpoons}} \; [\text{ArH}, \text{NO}_2\text{Y}] \qquad (9.41)$$

$$[\text{ArH}, \text{NO}_2\text{Y}] \; \underset{}{\overset{h\nu_{CT}}{\rightleftharpoons}} \; [\text{ArH}^{+\bullet}, \text{NO}_2\text{Y}^{-\bullet}] \qquad (9.42)$$

$$[\text{ArH}^{+\bullet}, \text{NO}_2\text{Y}^{-\bullet}] \; \longrightarrow \; [\text{ArH}^{+\bullet}, \text{NO}_2^{\bullet}] \, \text{Y}^- \qquad (9.43)$$

$$[\text{ArH}^{+\bullet}, \text{NO}_2^{\bullet}] \; \longrightarrow \; \text{Ar} \overset{+}{\underset{\text{NO}_2}{\overset{\displaystyle H}{<}}} \qquad (9.44)$$

$$\text{Ar} \overset{+}{\underset{\text{NO}_2}{\overset{\displaystyle H}{<}}} \; \longrightarrow \; \text{Ar} - \text{NO}_2 + \text{H}^+ \qquad (9.45)$$

Scheme 9.4

Let us now consider the key step in electrophilic aromatic substitution—formation of the Wheland intermediate [first step, Eq. (9.39)]—with the aid of the curve-crossing model. We will see that by building up a reaction profile for this process, the curve-crossing model provides a useful framework for relating the π complex, the σ complex, and the transition state, as well as describing the relationship between polar and ET pathways, and between ground- and excited-state pathways.

9.8.1 A Curve-Crossing Approach to Electrophilic Reactivity

In configuration terms the reactants and the σ complex may be represented by **9.21** and **9.22**, and designated DA and D^+A^-, respectively. The σ complex may be labeled D^+A^- since inspection shows that it is related to DA by a *single electron shift* (as are the S_N2 and other polar reactions of organic chemistry). Shifting an electron from the aromatic ring onto the electrophile E^+ sets up a configuration that describes the Wheland intermediate—an odd electron on the ring and one on the electrophile, which can couple up to form the new σ bond. These are the two configuration building blocks with which we can build up the reaction profile for the first step of electrophilic substitution—formation of the Wheland intermediate—and explore the mechanistic implications of this formulation.

9.21   9.22

The fact that the barrier for electrophilic aromatic substitution may be constructed from the avoided crossing of DA and D^+A^- configurations immediately tells us that electrophilic substitution and electron-transfer reactions are closely related, precisely in the way we discovered for nucleophilic substitution. The polar pathway is followed when the electron shift and Ar---E bond formation occur within a single step [Eq. (9.39)]. If, however, the electron-shift and bond-making process are decoupled, that is, they take place in two discrete steps, then the reaction pathway involves an initial electron transfer, after which the radical pair may combine to form the σ complex [Eq. (9.40)]. Thus, the form of the product configuration (**9.22**) as a charge-transfer configuration, emphasizes the intimate connection between the polar and electron-transfer pathways.

The way in which we can build up the ground-state energy profile for the polar pathway, leading to the formation of the Wheland intermediate, is illustrated in Fig. 9.9. As we have noted in many earlier configuration-mixing diagrams, reactant and product configurations DA and D^+A^-, illustrated by the two firm lines in Fig. 9.9, cross one another. The DA configuration, initially low in energy since it is a good descriptor of the reactants, rises in energy along the reaction coordinate. The D^+A^- configuration, on the other hand, is initially high in energy relative to DA since it is obtained by the vertical excitation (i.e., geometries remain frozen) of an electron from D to A. However, as one proceeds along the reaction coordinate, the energy of D^+A^- drops sharply due to coupling of the two odd electrons to form the new σ bond.

The ground- and excited-*state* energy curves for the formation of the σ complex are obtained by interaction of the two configuration curves. The two state curves are shown schematically by the dashed lines in Fig. 9.9; the lower dashed curve describes the ground-state energy profile while the upper curve describes an excited-state surface. As described previously (Chapter 4), the crossing of the state curves is avoided. From this diagram the electronic rela-

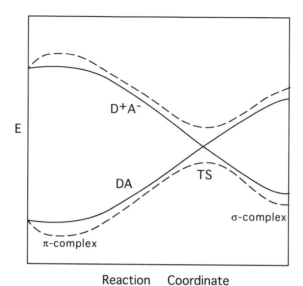

Fig. 9.9. Curve-crossing diagram showing the generation of ground- and excited-state energy profiles for the first step of electrophilic aromatic substitution from the configuration mixing of DA and D^+A^- curves (see text).

tionship between the π complex, the transition state leading to the formation of the σ complex, and the σ complex itself, becomes clearer.

The π complex between the arene and the electrophile is an example of a charge-transfer complex, and as such can be understood by considering the interaction of DA and D^+A^- configurations (see Section 7.3). As the donor and acceptor approach one another, some mixing of D^+A^- into DA takes place, leading to a reduction in energy. The wave function of this complex may therefore be written as

$$\pi \text{ complex} = DA + \lambda(D^+A^-) \qquad \lambda < 1 \qquad (9.46)$$

where the normalization constant is not shown for simplicity. Thus in electronic terms the charge-transfer complex is primarily described by DA, the low-energy configuration, with a small contribution of D^+A^-, the high-energy configuration. The location of the π complex is indicated in Fig. 9.9 by the local minimum near the reactant geometry in the lower dashed line.

How can we describe the transition state? Since the transition state is close to the crossing point of DA and D^+A^- (a point where the two configurations are of equal energy) the transition state wave function may be described by an *equal* contribution of the two configurations DA and D^+A^-, and therefore may be written as

$$TS = \frac{1}{\sqrt{2}} \{DA + D^+A^-\} \qquad (9.47)$$

Thus both the π complex and the reaction transition state may be described electronically by a mixing of the two interacting configurations, but in different proportions.

Finally, in configuration terms the σ complex would be described primarily by D^+A^-. The electron shift, necessary for the establishment of the new σ bond to form the Wheland intermediate, has taken place in full.

$$\sigma \text{ complex} = D^+A^- \tag{9.48}$$

Does the π complex lie along the reaction coordinate for σ-complex formation? Reference to Fig. 9.9 suggests that it does, though this point cannot be stated with certainty. Given that the reaction coordinate initially involves the approach of the substrate and the electrophile, followed by geometric changes that lower the energy of D^+A^- (so as to reduce the $DA - D^+A^-$ energy gap and increase $DA - D^+A^-$ mixing), it seems credible that the π complex would represent the beginning of the mixing process. However, there is no direct evidence that this is the case.

Note that the electronic description of the transition state as described in Eq. (9.47), applies to both the polar pathway [Eq. (9.39)] *and* the ground-state ET pathway [Eq. (9.40)]. In the case of the polar pathway, there is substantial bonding between D and A in the transition state, while in the ET pathway the extent of bonding would depend on whether the electron transfer is outer sphere or inner sphere. If the ET mechanism is via an outer-sphere pathway, then there is effectively no bonding between D and A in the transition state, while for an inner-sphere pathway some bonding interaction between D and A does occur. In that sense, the outer-sphere pathway should be viewed as representing one extreme of a mechanistic continuum, in which the bonding interaction in the transition state may vary from very slight to substantial.[19]

What factors will determine whether the thermal reaction pathway will follow the polar concerted mechanism or the stepwise ET mechanism? As described in detail in Section 9.5, treating the profile of polar reactions as being generated from the avoided crossing of DA and D^+A^- configurations provides us with a means of qualitatively predicting the factors that are likely to favor one pathway over the other. In summary, the important factors are (a) the donor–acceptor pair ability, (b) the strength of the newly formed bond between D and A, and, (c) steric factors that may impede D–A approach. *An electron-transfer mechanism will be encouraged when a good donor–acceptor pair is involved, when the newly formed bond is weak, and when steric factors impede the formation of the tight transition state that is required in the polar pathway.* In essence, *enthalpic* factors that favor the polar pathway (through a tight transition state) compete with *entropic* factors that favor the ET pathway (through a loose transition state).

The relationship between the thermal and photochemical pathways for electrophilic aromatic substitution also becomes clearer by considering a curve-crossing diagram, and this is illustrated in Fig. 9.10. The configuration curves

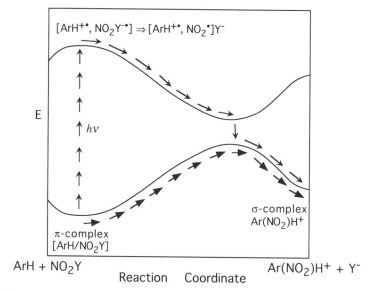

Fig. 9.10. Energy diagram showing thermal (bold arrows) and photochemical (light arrows) pathways for the formation of $Ar(NO_2)H^+$ during electrophilic aromatic nitration (see text).

of Fig. 9.9 are now deleted and just the state curves, obtained through the interaction of the configuration curves, are shown. Thermal and photochemical pathways are indicated by the two sets of arrows.[a]

According to Scheme 9.4 the photochemical pathway (indicated by the light arrows) is initiated by the excitation of the charge-transfer complex from the ground surface to the excited surface [Eq. (9.42)]. This excitation effectively involves the vertical transfer of an electron from the arene to the electrophile. In the case of nitration this generates the radical ion pair $[ArH^{+\cdot}, NO_2Y^{-\cdot}]$, where NO_2Y is the nitrating agent. Fragmentation of the nitrating agent radical anion then yields $NO_2^\cdot$ within the complex $[ArH^{+\cdot}, NO_2^\cdot]Y^-$ [Eq. (9.43)]. The radical species within this complex can then couple up to generate the σ complex [Eq. (9.44)] and so return to the ground-state surface. Thus, electron transfer is a key step in the photochemical pathway. Of course the σ complex may also be generated along the ground-state (thermal) surface (indicated by the bold arrows in Fig. 9.10).

At the time of this writing, the mechanism of thermal nitration of aromatics

[a]It is worth pointing out that a detailed understanding of the photochemical pathway is still lacking. Shaik and Reddy[20] pointed out that in addition to "funnels," which can provide a channel for radiationless decay from the excited-state to the ground-state surface, decay may proceed by following a different reaction coordinate, in which the "funnel" (which bridges across the avoided crossing region) is converted to an *actual* crossing. These mechanistic details are still to be resolved.

(along the ground-state energy profile) remains controversial. Considerable debate centers on the question of whether thermal nitration proceeds via a stepwise ET pathway [Eq. (9.40)], or via direct formation of the σ complex [Eq. (9.39)]. The source of the controversy stems from conflicting experimental results. In 1977 Perrin[21] focused attention on the possibility of an ET pathway when he reported that the ratio of 1-nitronaphthalene and 2-nitronaphthalene obtained from the nitration of naphthalene in acetonitrile

$$
\text{naphthalene} + NO_2^+ \longrightarrow \text{1-nitronaphthalene} + \text{2-nitronaphthalene}
$$

$$
9.2 \quad : \quad 1
$$

$$
(9.49)
$$

was the same (within experimental error) as that produced through reaction of the electrochemically produced naphthalene radical cation and neutral NO_2

$$
[\text{naphthalene}]^{+\bullet} + {}\bullet NO_2 \longrightarrow \text{1-nitronaphthalene} + \text{2-nitronaphthalene}
$$

$$
10.9 \quad : \quad 1
$$

$$
(9.50)
$$

Perrin interpreted the similar regioselectivity to mean that the two reactions share a common product-determining step, that is, the mechanism of electrophilic substitution in this case proceeds via an electron-transfer step to yield the naphthalene radical cation and neutral NO_2, which then combine to give reaction products [as by Eq. (9.40)]. This interpretation is consistent with Kochi's photochemical and thermal nitration results, which suggested similar product distributions from photochemical and thermal nitration.

However, subsequent work by Eberson et al.[22] questioned these experimental results. Naphthalene nitration experiments in dichloromethane by Eberson and Radner[23] led to different product ratios [15:1 for the regular nitration reaction, Eq. (9.49), and 40:1 for the radical combination reaction, Eq. (9.50)], so that the mechanistic significance of Perrin's earlier results is questioned.[22] Eberson also found that the actual mechanistic pathway followed during the photochemical nitration reaction of naphthalene by tetranitromethane involves formation of adducts of the kind

9.23

These adducts are formed from the collapse of the [ArH$^{+\bullet}$, NO$_2^{\bullet}$, C(NO$_2$)$_3^-$] triad following the photoexcitation, which then decompose to generate nitro-naphthalene products. Eberson finds that the naphthalene nitration product distribution is strongly influenced by the decomposition of these adducts and so questions the conclusions drawn from Kochi's product distribution studies, which suggested that photochemical and thermal product distributions are the same. Eberson also argued that due to its high reorganization energy, the NO$_2^+$ ion may not be a sufficiently strong one-electron oxidant to be able to oxidize most aromatic substrates. Thus, Eberson's results have undermined the case for electron transfer during thermal nitration.[22]

Clearly, the mechanistic details of aromatic nitration, both thermal and photochemical are more complex than originally thought, and will require further study. However, even at this early juncture, it is apparent that an electron-transfer perspective on this most basic of organic reactions (which, significantly, is employed by *both* sides of this debate) is leading to a general reassessment of the mechanism of electrophilic aromatic substitution as a whole.

In summary, a curve-crossing analysis of electrophilic aromatic substitution emphasizes the intimate connection between polar and electron-transfer mechanisms as well as the photochemical pathway. The essence of an electrophilic substitution reaction is a single-electron shift from the arene to the electrophile. If the electron shift is *coupled* to σ-bond formation then the reaction corresponds to the conventional polar pathway that proceeds via the Wheland intermediate. If, however, the electron-transfer and bond coupling processes are *uncoupled*, as they are for good donor–acceptor pairs, or in the photochemical process, then reaction may proceed via electron transfer and actual radical intermediates will be formed. The precise mechanistic pathway that is followed in different cases, though controversial and hotly debated at the present time, is likely to be ultimately formulated using electron-transfer terminology.

REFERENCES

1. H. Taube, *Electron Transfer Reactions of Complex Ions in Solution*, Academic Press, London, 1970.

2. T. Lund and H. Lund, *Acta Chem. Scand., Ser. B*, **40**, 470 (1986). See also T. Lund and H. Lund, *Acta Chem. Scand., Ser. B*, **41**, 93 (1987); **42**, 269 (1988).

3. C. D. Ritchie, *Acc. Chem. Res.* **5**, 348 (1972).

4. C. G. Swain and C. B. Scott, *J. Am. Chem. Soc.* **75**, 141 (1953).

5. R. G. Pearson, *J. Org Chem.* **52**, 2131 (1987).

6. F. G. Bordwell, T. A. Cripe, and D. L. Hughes. In *Nucleophilicity*, ACS Advances in Chemistry Series No. 215, J. M. Harris and S. P. McManus, Eds. American Chemical Society, Washington, DC, 1987.

7. For a more quantitative application of the curve-crossing model to S_N2 reactions, see S. S. Shaik, H. B. Schlegel, and S. Wolfe, *Theoretical Aspects of Physical Organic Chemistry. The S_N2 Mechanism*, Wiley-Interscience, New York, 1992.

8. A. J. Parker, *Chem. Rev.* **69**, 1 (1969).

9. M. J. Pellerite and J. I. Brauman, *J. Am. Chem. Soc.* **102**, 5993 (1980); M. J. Pellerite and J. I. Brauman, *J. Am. Chem. Soc.* **105**, 2672 (1983); J. A. Dodd and J. I. Brauman, *J. Am. Chem. Soc.* **106**, 5356 (1984).

10. (a) A. Pross, *Acc. Chem. Res.* **18**, 212 (1985); (b) E. C. Ashby, *Acc. Chem. Res.* **21**, 414 (1988); (c) J.-M. Saveant, *Adv. Phys. Org. Chem.* **26**, 1 (1990);

11. E. B. Troughton, K. E. Molter, and E. M. Arnett, *J. Am. Chem. Soc.* **106**, 6726 (1984).

12. (a) J. F. Bunnett, *Acc. Chem. Res.* **11**, 413 (1978); (b) R. A. Rossi, *Acc. Chem. Res.* **15**, 164 (1982); (c) J.-M. Saveant, *J. Phys. Chem.* **98**, 3716 (1994).

13. (a) Z. Rappoport, *Acc. Chem. Res.* **25**, 474 (1992); (b) M. Ochiai, K. Oshima, and Y. Masaki, *J. Am. Chem. Soc.* **113**, 7059 (1991); (c) M. Gloukhovtsev, A. Pross, and L. Radom, *J. Am. Chem. Soc.* **116**, 5961 (1994); (d) Z. Rappoport and B. A. Shainyan, *J. Org. Chem.* **58**, 3421 (1993);

14. B. L. Knier and W. P. Jencks, *J. Am. Chem. Soc.* **102**, 6789 (1980).

15. D. J. McLennan and A. Pross, *J. Chem. Soc. Perkin Trans. 2*, 981 (1984).

16. A. Pross, K. Aviram, R. C. Klix, D. Kost, and R. D. Bach, *Nouv. J. Chim.* **8**, 711 (1984).

17. P. R. Young and W. P. Jencks, *J. Am. Chem. Soc.* **101**, 3288 (1979).

18. J. K. Kochi, *Acc. Chem. Res.* **25**, 39 (1992); *Adv. Phys. Org. Chem.* **29**, 185 (1994).

19. L. Eberson and S. S. Shaik, *J. Am. Chem. Soc.* **112**, 4484 (1990).

20. S. Shaik and C. A. Reddy, *J. Chem. Soc. Faraday Trans.* **90**, 1631 (1994).

21. C. L. Perrin, *J. Am. Chem. Soc.* **99**, 5516 (1977).

22. L. Eberson, M. P. Hartshorn, and F. Radner. In *Advances in Carbocation Chemistry*, Vol. 2, J. M. Coxon, Ed. JAI Press, London, 1995.

23. L. Eberson and F. Radner, *Acta Chem. Scand.* **B 34**, 739 (1980).

10

RADICAL AND PERICYCLIC REACTIVITY

Radical and pericyclic reactions share certain theoretical characteristics that distinguish them from polar reactions. In this chapter we survey basic features of these two reaction types and the factors that govern their reactivity, utilizing the curve-crossing model for much of the analysis.

10.1 RADICAL REACTIVITY

Radical reactions are generally multistep processes whose individual steps can be classified as (a) initiation, (b) propagation, or (c) termination steps. The initiation step, as its name suggests, initiates the reaction sequence. It is the step in which a *radical* intermediate is formed from *nonradical* precursors. This step involves bond homolysis, which may be induced either thermally or photochemically. Two common radical precursors are benzoyl peroxide, which contains the weak $O-O$ bond, making it relatively easy to homolyze

$$C_6H_5-\underset{\underset{O}{\|}}{C}-O-O-\underset{\underset{O}{\|}}{C}-C_6H_5 \quad\xrightarrow{\Delta}\quad 2C_6H_5-\underset{\underset{O}{\|}}{C}-O^{\bullet} \quad (10.1)$$

and 2,2'-azo-bisisobutyronitrile, whose decomposition

$$CH_3-\underset{\underset{CN}{\|}}{\overset{\overset{CH_3}{\|}}{C}}-N=N-\underset{\underset{CN}{\|}}{\overset{\overset{CH_3}{\|}}{C}}-CH_3 \quad\xrightarrow{\Delta}\quad 2CH_3-\underset{\underset{CN}{\|}}{\overset{\overset{CH_3}{\|}}{C^{\bullet}}} + N_2 \quad (10.2)$$

is facilitated by the high stability of the nitrogen molecule that is generated. Thus initiation reactions are characterized by the relative ease with which they produce radicals.

The propagation steps take place once radicals have been generated in the initiation step. These steps involve a reaction between the radical and a reactant molecule and lead either to the direct formation of the reaction product, or, more typically, to the formation of product precursors, which react further to yield the reaction products. Propagation steps belong to one of two basic types: addition of the initial radical to a double bond

$$R\cdot + C{=}C \longrightarrow \overset{\displaystyle R}{\underset{\displaystyle}{\diagdown}}C{-}C\cdot \qquad (10.3)$$

or abstraction of an atom or group from another molecule

$$R\cdot + R'{-}X \rightarrow R{-}X + R'\cdot \qquad (10.4)$$

Both lead to the generation of new radical species that may then continue to react in further propagation reactions until two radicals combine in a termination step

$$R\cdot + R'\cdot \rightarrow R{-}R' \qquad (10.5)$$

Thus termination steps do the opposite of the initiation steps. They eliminate radicals from the reaction mixture, and by doing so inhibit the radical process, and ultimately bring it to an end.

Of the three types of reactions, the group of greatest chemical interest is the class of propagation reactions. It is in the propagation steps that the "real" chemistry occurs, and it is the rates of the propagation steps that are particularly important in governing the overall rate of radical processes.

What factors govern the rates of propagation reactions, such as those represented in Eqs. (10.3) and (10.4)? By using the curve-crossing model we can build up the reaction profile for a radical addition or abstraction process, and thereby identify those factors that govern reactivity in these fundamental bond-making process. Let us do this for the first step of a radical addition reaction [Eq. (10.3)].

10.2 A CURVE-CROSSING APPROACH TO RADICAL ADDITION

The first step in building up a reaction profile for the addition of a carbon radical to an alkene [Eq. (10.3)] is to specify the VB configurations from which the profile is constructed using the DA formalism (where the radical is arbitrarily labeled D, and the alkene is labeled A). The four key configurations are shown in **10.1–10.4** and are designated by the labels DA, D^3A^*, D^+A^-, and

D^-A^+. These four configurations are obtained by rearranging the three valence electrons—the odd electron on the radical and the two electrons in the π bond—since it is these three electrons that reorganize during the course of the reaction. Let us now describe this set of VB configurations in some detail.[1]

$$
\begin{array}{cc}
\overset{\displaystyle C\uparrow}{\underset{\displaystyle \cdot \overset{\uparrow\downarrow}{-}\, \cdot}{}} & \overset{\displaystyle C\downarrow}{\underset{\displaystyle \cdot \overset{\uparrow\uparrow}{-}\, \cdot}{}} \\[4pt]
\text{DA} & \text{D}^3\text{A*} \\[2pt]
\mathbf{10.1} & \mathbf{10.2}
\end{array}
$$

$$
\begin{array}{cc}
C^+ \qquad\qquad C^+ & C^- \qquad\qquad C^- \\
\dot{C}-\ddot{C}^- \;\leftrightarrow\; {}^-\ddot{C}-\dot{C} & \dot{C}-C^+ \;\leftrightarrow\; {}^+C-\dot{C} \\[4pt]
D^+A^- & D^-A^+ \\[2pt]
\mathbf{10.3} & \mathbf{10.4}
\end{array}
$$

The reactant configuration DA (**10.1**) is a good approximation for the electronic distribution in the reactants; the odd electron is on the carbon radical and the ethylene molecule is in a singlet configuration. In similar fashion the product configuration D^3A* (**10.2**) is a good representation for the electronic distribution in the products; one electron is still associated with each of the three carbon atoms as in DA. However, the spin arrangement has been changed so that the spin pairing is between the two electrons of the new C—C σ bond and *not* between the two electrons of the π bond. The product configuration is designated D^3A* because it is obtained from the reactant configuration by the excitation of the π bond from the singlet to the triplet configuration (i.e., A to $^3A*$).

Additional configurations are the charge-transfer configurations, labeled D^+A^- (**10.3**) and D^-A^+ (**10.4**). Configuration D^+A^- is obtained by the transfer of an electron from the radical to the alkene, while D^-A^+ is obtained by the transfer of an electron from the alkene to the radical. Since both D^+A^- and D^-A^+ are delocalized in nature, each is represented by two VB configurations.

An energy profile for the addition of a radical to a double bond may now be built up from these configurations by plotting each of them as a function of the reaction coordinate. This is shown schematically in Fig. 10.1.[1e,f] The energetics shown in the diagram approximately represent those for addition of the methyl radical to ethylene. The reactant configuration (DA) rises in energy as we move along the reaction coordinate because repulsive interactions set in between the odd electron on the radical and the two π electrons; in this configuration no bonding can take place between the radical and an alkene carbon atom because the electron spins do not allow it.

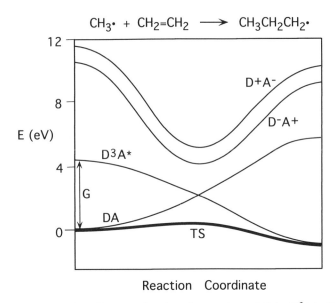

Fig. 10.1. Curve-crossing diagram showing the mixing of DA, D^3A^*, D^+A^-, and D^-A^+ configurations in the schematic generation of the ground-state reaction surface (bold line) for the addition of an alkyl radical to monosubstituted ethylenes. The curves are drawn to correspond approximately to the energies (where available) for the system of methyl radical plus ethylene.

The product configuration D^3A^* behaves in reverse. At the reactant geometry it is high in energy (4.3 eV—the vertical triplet excitation energy of ethylene; see Table 7.8). However, if we move along the reaction coordinate the energy drops sharply, since the odd electron on the radical can now pair up with an uncoupled electron in the π system to form the new C—C bond. Thus even though D^3A^* is an *excited* configuration in the reactant geometry, it is a *ground* configuration in the product geometry. At the product end of the reaction coordinate, the energy of D^3A^* for methyl adding to ethylene actually drops below the initial energy of DA reflecting the fact that the reaction is somewhat exothermic (by ~ 22.3 kcal mol^{-1}).

Now let us consider D^+A^- and D^-A^+ configurations and see how their energies vary along the reaction coordinate. At infinite separation the energies of these two configurations may be obtained from ionization potential and electron affinity data. For the system of methyl radical and ethylene these configurations are initially high in energy (a recent theoretical estimate places them at 11.63 and 10.54 eV, respectively).[1e,f] However, due to the electrostatic attraction between positive and negative moieties, both will drop sharply in energy as reactants approach one another. A simple electrostatic estimate based on a separation of 2.4 Å between positive and negative centers suggests this drop will be about 6 eV [see Eq. (4.3) in Section 4.3.2]. A schematic representation of the energies of D^+A^- and D^-A^+ for methyl radical adding to ethylene is shown in Fig. 10.1.

We can now see why the addition of a radical to a double bond has an activation barrier. The electronic configuration of the system, initially described by DA, must reorganize to that in D^3A^*. How does this come about? It occurs through the geometric distortions that *raise* the energy of DA and simultaneously *lower* the energy of D^3A^*. As the system moves progressively further along the reaction coordinate, these two configurations increasingly mix with one another. At the crossing point the two configurations mix equally since they are now isoenergetic, and this mixing leads to their *avoided crossing*. Thus it is the avoided crossing of DA and D^3A^* that is primarily responsible for generating the activation barrier in the addition reaction of a radical to a double bond. The *ground-state energy curve* for addition of a radical to an alkene is shown by the bold line in Fig. 10.1, with the transition state, labeled TS, the highest point on that curve. Due to the scale of the energy coordinate the calculated barrier height for this reaction (~ 9.3 kcal mol^{-1} or 0.4 eV) appears to be very small.

How do the charge-transfer configurations affect the TS description? If the TS is described by just an equal mixing of DA and D^3A^*, then the TS wave function may be written

$$\Psi_{TS} \cong \frac{1}{\sqrt{2}} [DA + D^3A^*] \tag{10.6}$$

If, however, we allow for the mixing in of one of the charge-transfer configurations, then Eq. (10.6) becomes

$$\Psi_{TS} \cong N \left\{ \frac{1}{\sqrt{2}} [DA + D^3A^*] + \lambda \Phi_{CT} \right\} \tag{10.7}$$

where Φ_{CT} is either D^+A^- or D^-A^+ and mixes into the TS with a mixing coefficient λ, and N is a normalization factor, whose significance is mathematical rather than chemical. Thus the essence of Eq. (10.7) may be expressed qualitatively by

$$\Psi_{TS} = DA \leftrightarrow D^3A^* \leftrightarrow D^+A^- \leftrightarrow D^-A^+ \tag{10.8}$$

that is, the transition state may be described as a resonance hybrid of reactant, product, and charge-transfer configurations. In practice, the possible involvement of charge-transfer configurations will largely depend on their energies so their contribution is likely to be quite variable, ranging from negligible when their energies are relatively high, to substantial when they are close in energy to the crossing point of reactant and product configurations. Mixing in of a charge-transfer configuration, when it does take place, will have two important consequences. First, it will lead to stabilization of the transition state, and second, the transition state will take on some of the charge-transfer character of that configuration.

10.2.1 Factors Influencing the Barrier Height in Radical Addition

Building up the reaction profile from configuration building blocks, as illustrated in Fig. 10.1, enables us to specify several of the factors that are likely to influence the barrier height. In Section 4.4 we described the general features of the curve-crossing model and pointed out the factors that are likely to affect the barrier height. For radical addition these can be summarized as

1. *Reaction Exothermicity.* Stabilizing the reaction products relative to reactants (i.e., increasing reaction exothermicity) is predicted to lead to a reduction in the barrier height. (This is just a manifestation of the rate–equilibrium relationship, described in Section 6.5.)

2. *The Initial Energy Gap G.* In Section 4.4.2 we pointed out that the initial energy gap is an important factor in governing barrier heights. For the case of a radical addition reaction the initial energy gap is between DA and D^3A^* configurations (illustrated in Fig. 10.1 and defined by the alkene vertical singlet–triplet excitation energy, labeled G). The curve-crossing model predicts that *decreasing* G will decrease the barrier height. Note that the effect of G on reactivity is an example of a *nonthermodynamic* contribution to reactivity.

While the effect of substituents on reaction exothermicity and on the singlet–triplet gap might appear to be independent of one another, it turns out that for radical addition reactions they are closely related. Consider the addition of a radical $R\cdot$ to a family of monosubstituted alkenes

$$R\cdot\ +\ CH_2\!=\!CHX\ \rightarrow\ RCH_2CHX\cdot \tag{10.9}$$

Given that an unpaired electron on C is located adjacent to the substituent in *both* the product radical $RCH_2CHX\cdot$ (**10.5**) *and* the triplet alkene $CH_2\!=\!CHX$ (**10.6**), the effects of substituents on the energies of these two species are likely to be similar. *In other words, changes in the singlet–triplet excitation energies are expected to follow changes in the reaction exothermicities.* The consequence of this linkage is that substituent effects on these two parameters will affect the energy of the product configuration, as illustrated in Fig. 10.2. Thus a substituent that increases reaction exothermicity (lowering Z to Z$'$) will also reduce the initial energy gap (lowering Y to Y$'$) so that the *entire* D^3A^* surface moves to lower energy. As a result the energy of the transition state (approximated by the crossing point of DA and D^3A^*) drops from TS to TS$'$ and its *structure* moves to a position *earlier* along the reaction coordinate. This type of response to substituent effects is expected to lead to a *barrier height–enthalpy relationship:* the energy form of the rate–equilibrium relationship.

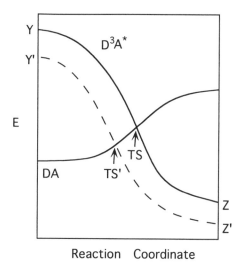

Reaction   Coordinate

Fig. 10.2. Schematic energy diagram showing the effect of a substituent that lowers the energy of the product configuration D^3A^*, at both reactant and product ends of the reaction coordinate (Y drops to Y' and Z drops to Z'). The transition state moves from TS to TS'.

$$\overset{\displaystyle\uparrow}{RCH_2\overset{\bullet}{C}HX} \qquad \overset{\displaystyle\uparrow\quad\uparrow}{\overset{\bullet}{C}H_2—\overset{\bullet}{C}HX}$$

10.5   **10.6**

Experimental and computational data confirm the link between barrier height and reaction enthalpy.[1e, f] The computed barrier height and enthalpy data for methyl radical addition to substituted alkenes [Eq. (10.9), where R = CH_3], listed in Table 10.1, are found to correlate well for a wide range of substituents. This is illustrated by the correlation line in Fig. 10.3. However, when points for a second radical ($\cdot CH_2OH$) are added to the graph, these are found to lie *below* the $\cdot CH_3$ correlation line (Fig. 10.3). The same is true when the points for the $\cdot CH_2CN$ radical are added to the $\cdot CH_3$ plot, as shown in Fig. 10.4. Here the points are also found to lie below the $\cdot CH_3$ correlation line. Thus a barrier–enthalpy plot does *not* encompass a number of different radicals. The reason the correlation fails when different radicals, such as $\cdot CH_2OH$ and $\cdot CH_2CN$ are considered, can be understood by considering the role of charge-transfer configurations.

3. *Stability of Charge-Transfer Configurations.* The curve-crossing model predicts that when D^+A^- or D^-A^+ are low in energy they will mix into the transition state wave function and lower the energy of the transition state, thereby leading to enhanced reactivity.

TABLE 10.1. Calculated Barrier Heights and Enthalpies for Addition of $\cdot CH_3$, $\cdot CH_2OH$, and $\cdot CH_2CN$ to $CH_2{=}CHX^a$

X	Barrier (kcal mol^{-1})			Enthalpy (kcal mol^{-1})		
	$\cdot CH_3$	$\cdot CH_2OH$	$\cdot CH_2CN$	$\cdot CH_3$	$\cdot CH_2OH$	$\cdot CH_2CN$
F	9.5	8.4	10.1	−22.5	−20.9	−15.1
H	9.3	7.8	10.2	−22.3	−20.8	−15.1
NH$_2$	8.7	7.8	7.3	−23.9	−21.8	−17.2
Cl	7.8	5.9	8.6	−25.3	−23.4	−17.8
CHO	6.9	4.4	8.1	−28.8	−28.3	−22.2
CN	5.8	2.8	7.8	−30.9	−29.6	−22.3

^{a}Values obtained using the QCISD/6-311G** basis set and incorporating zero-point vibration-energy corrections. Data taken from M. W. Wong, A. Pross, and L. Radom, *J. Am. Chem. Soc.* **116**, 6284 (1994).

For methyl radical, the energies of the charge-transfer configurations, D^+A^- and D^-A^+, with different alkenes $CH_2{=}CHX$ (listed in Table 10.2) are relatively high (normally above 9.5 eV). This means that methyl is generally neither a good electron donor, nor a good electron acceptor. Accordingly, *the methyl radical does not normally exhibit well-defined polar character toward*

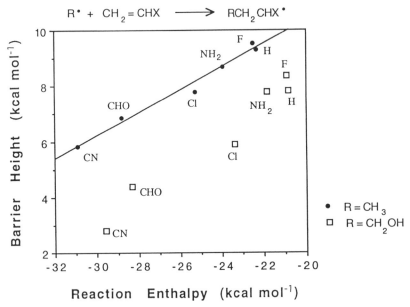

Fig. 10.3. Plot of the computed ab initio barrier height against reaction enthalpy (kcal mol^{-1}, calculated at the QCISD/6-311G** + ZPE level) for the addition of $\cdot CH_3$ (full circles) and $\cdot CH_2OH$ (empty squares) to alkenes $CH_2{=}CHX$ (X = H, NH$_2$, F, Cl, CHO, and CN).

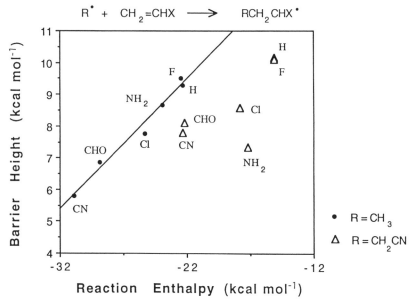

Fig. 10.4. Plot of computed ab initio barrier height against reaction enthalpy (kcal mol^{-1}, calculated at the QCISD/6-311G** + ZPE level) for the addition of ·CH$_3$ (full circles) and ·CH$_2$CN (empty triangles) to alkenes CH$_2$=CHX (X = H, NH$_2$, F, Cl, CHO, and CN).

alkenes. In the absence of significant polar effects, a barrier height–enthalpy correlation is observed, as described by the correlation line in Fig. 10.3.

For ·CH$_2$OH and ·CH$_2$CN the situation is quite different; for these two radicals, one of the two charge-transfer configurations is relatively low in energy (below 9.5 eV) and this means that these two radicals are likely to exhibit polar character. For ·CH$_2$OH, D$^+$A$^-$ is generally the low-energy configuration so ·CH$_2$OH will tend to exhibit *nucleophilic* character. For ·CH$_2$CN, D$^-$A$^+$

TABLE 10.2. Calculated Charge-Transfer States (D$^+$A$^-$ and D$^-$A$^+$) Related to ·CH$_3$, ·CH$_2$OH, and ·CH$_2$CN Addition to CH$_2$=CHX^{a,b}

	·CH$_3$		·CH$_2$OH		·CH$_2$CN	
X	D$^+$A$^-$	D$^-$A$^+$	D$^+$A$^-$	D$^-$A$^+$	D$^+$A$^-$	D$^-$A$^+$
F	11.39	10.33	9.05	10.51	11.78	8.78
H	11.64	10.54	9.30	10.72	12.03	8.99
NH$_2$	11.69	8.14	9.35	8.32	12.08	6.59
Cl	11.05	9.94	8.71	10.12	11.44	8.38
CHO	9.74	10.17	7.46	10.35	10.13	8.62
CN	10.00	10.94	7.66	11.12	10.39	9.39

^{a}Data taken from M. W. Wong, A. Pross, and L. Radom, *J. Am. Chem. Soc.* **116**, 6284 (1994).
^{b}In electronvolts (eV).

is the low-energy configuration, so $\cdot CH_2CN$ will tend to exhibit *electrophilic* character.

The energetic consequences of the polar character that these two radicals take on during the reaction can be observed by inspection of the barrier height–enthalpy correlations illustrated in Figs. 10.3 and 10.4. As we noted earlier, the points for $\cdot CH_2OH$ and $\cdot CH_2CN$ fall *below* the $\cdot CH_3$ correlation line. Or stated differently, for a given thermodynamic driving force, $\cdot CH_2OH$ and $\cdot CH_2CN$ have *lower* reaction barriers than $\cdot CH_3$. It appears that for these two polar radicals, charge-transfer contributions to the electronic structure of the transition state lead to a *lowering* of the transition state energy compared to that for $\cdot CH_3$, which is largely devoid of polar character. In fact, the deviation of the $\cdot CH_2OH$ and $\cdot CH_2CN$ points from the $\cdot CH_3$ correlation line may be used to estimate the *extent* of polar stabilization in the transition state for these two radicals. Thus for $\cdot CH_2OH$, the points for CN and CHO substituents appear to lie some 2.5–3.5 kcal mol^{-1} below the $\cdot CH_3$ line suggesting that these values reflect the extent of D^+A^- stabilization for the $\cdot CH_2OH/CH_2=CHCN$ and $\cdot CH_2OH/CH_2=CHCHO$ pairs. For the $\cdot CH_2CN$ radical, the point that deviates the most from the $\cdot CH_3$ correlation line is that for NH_2 (by ~ 3.5 kcal mol^{-1}) suggesting this to be the extent of transition state stabilization due to D^-A^+ mixing for the $\cdot CH_2CN/CH_2=CHNH_2$ pair.

In summary, we find that reaction enthalpy (and the singlet–triplet energy gap) as reflected in the energy of the D^3A^* configuration, is a key factor in governing the barrier height for radical addition to alkenes; increasing reaction exothermicity tends to decrease the reaction barrier. However, in those cases where charge-transfer configurations are relatively stable and close in energy to the crossing point of DA and D^3A^*, polar character in the transition state also becomes important. In such cases the transition state is further stabilized, leading to a lower reaction barrier and a breakdown in the barrier height–enthalpy correlation.

10.2.2   Regioselectivity in Radical Addition

The addition of a radical to a substituted alkene

$$R\cdot + H_2C=CHX \rightarrow RCH_2-CHX\cdot + \cdot CH_2-CHXR \qquad (10.10)$$

$$A \qquad\qquad\qquad B$$

can lead to two different products, A and B, depending on which carbon is attacked by the incoming radical. In most cases the radical attacks the least substituted carbon to give product A, however, in some cases regioselectivity is inverted and product B is formed preferentially. What factors govern regioselectivity in radical addition?[1d]

In Section 10.2.1 we demonstrated that reaction exothermicity is an important factor in governing barrier heights in radical addition. We would therefore expect that the relative stability of A and B will play an important role in determining the preferred product. This is certainly the case. In general, for the addition of a radical to a monosubstituted alkene, product A, where the radical center is stabilized by the substituent X, is more stable than product B, where interaction of the radical center with X cannot occur. However, in some instances this stability order reverses. For example, in the addition of $\cdot CH_3$ radical to $CH_2{=}CHF$ and $CH_2{=}CF_2$, MO calculations suggest that product A is the *less* stable product thermodynamically, yet it is still the one that is preferred kinetically. What is the reason for the preferential formation of the *less stable* isomeric product?

An obvious explanation that might account for the formation of the less stable product is the possible influence of steric factors. Given that the substituted end of the alkene would presumably be more sensitive to steric interactions than the unsubstituted (or less substituted) end, the preferential formation of product A, even though it may be less stable, can be explained. This explanation, however, is considered unlikely since the contribution of steric factors to regioselectivity are thought to be small. Radical addition to alkenes are in large part exergonic reactions, and so are expected to have *early* transition states with a long radical–carbon incipient bond (in fact MO calculations on carbon radicals suggest a transition state C---C bond length of about 2.2 Å). So what factor could be responsible for reversing the regioselectivity in certain cases?

A factor that appears to be important in influencing regioselectivity is the relative *spin density* in the triplet pair of the D^3A^* configuration.[1d] Consider the triplet state of a monosubstituted alkene ($CH_2{=}CHX$), where X is an electron-donor substituent. The electronic structure of this triplet may be represented by the mixing of three VB configurations (**10.7–10.9**). Structure **10.7**

$$\ddot{X}{-}\overset{\uparrow}{\underset{\cdot}{C}}{-}\overset{\uparrow}{\underset{\cdot}{C}} \;\longleftrightarrow\; {}^{+}\overset{\cdot}{X}{-}\overset{\uparrow}{\underset{\overset{=}{\cdot}}{C}}{-}\overset{\uparrow}{\underset{\cdot}{C}} \;\longleftrightarrow\; {}^{+}\overset{\cdot}{X}{-}\overset{\uparrow}{\underset{\cdot}{C}}{-}\overset{\uparrow}{\underset{\cdot\cdot}{C}}{}^{-}$$

$$\textbf{10.7} \qquad\qquad \textbf{10.8} \qquad\qquad \textbf{10.9}$$

is the dominant Lewis form; it has no charge separation but does suffer from triplet repulsion (since the unpaired electrons are on adjacent carbon atoms). Structure **10.8** relieves the triplet repulsion present in **10.7** by locating the unpaired electrons on atoms that are no longer adjacent, but replaces it with charge separation, which is unfavorable. Structure **10.9** suffers from both triplet repulsion *and* large charge separation (the charge separation distance is greater in **10.9** than in **10.8**), and so is less stable than **10.8**. Hence, **10.8** will be the secondary VB configuration that mixes in with the main configuration **10.7**, and as a result *the spin density on the two carbon atoms in the triplet alkene* *will not be equal, but will be higher on the β carbon* as indicated by

10.10

A similar argument for an electron-accepting group X attached to a triplet pair leads to the same conclusion; *the spin density will be larger on the β carbon.*

Can this difference in spin density on the two alkene carbon atoms in the ground state of the triplet alkene affect regioselectivity? Recall, that the reason D^3A^* drops in energy as the radical approaches the alkene is that the odd electron on the radical and one of the triplet pair electrons couple up to form a bond pair. The larger the spin density on the carbon that undergoes this bond formation, the greater the stabilization energy will be, and this will reflect itself in a steeply descending D^3A^* curve and a reduced barrier (see Fig. 4.9, which shows the effect of curve slopes on barrier height). So a *larger* spin density on the β carbon may lead to a *kinetic* preference for attack at that carbon, even though this isomer may be the less thermodynamically stable one.

In summary, it appears that regioselectivity in radical addition is governed by two main factors: triplet spin density (a kinetic factor) and product stability (a thermodynamic factor). In most radical addition reactions these two effects are reinforcing, so product A is normally preferred. However, when these two effects oppose one another, inversion of regioselectivity may be observed.

10.3 RADICAL SUBSTITUTION

Radical substitution reactions, written in general form as

$$X\cdot + Y-Z \rightarrow X-Y + Z\cdot \qquad (10.11)$$

involve the abstraction of an atom or group Y from a molecule by some radical species. A particularly common reaction of this type is when an H atom is abstracted

$$X\cdot + H-Z \rightarrow X-H + Z\cdot \qquad (10.12)$$

In Section 10.2 we described the VB configurations that are used to build up the reaction profile for radical addition. It turns out that the energy profile for radical substitution is built up from precisely the same configurations, namely, DA, D^3A^*, D^+A^-, and D^-A^+. These are depicted in **10.11–10.14.**

$$\overset{\uparrow}{X\cdot} \ \overset{\uparrow\uparrow}{Y\cdot} \ \overset{\downarrow}{\cdot Z} \qquad \overset{\uparrow\uparrow}{X\cdot} \ \overset{\downarrow}{\cdot Y} \ \overset{\downarrow}{\cdot Z} \qquad X^+ \ (Y \overset{\cdot}{-} Z)^- \qquad X:^- \ (Y \cdot Z)^+$$

$$DA \qquad\qquad D^3A^* \qquad\qquad D^+A^- \qquad\qquad D^-A^+$$

$$\textbf{10.11} \qquad\qquad \textbf{10.12} \qquad\qquad \textbf{10.13} \qquad\qquad \textbf{10.14}$$

Thus, by analogy with radical addition, the factors that are expected to govern radical substitution are (a) the reaction enthalpy and (b) the magnitudes of the charge-transfer parameters I and A, for the radical and the substrate. These expectations have been confirmed both experimentally and computationally. Hydrogen atom abstraction is found to be sensitive to the strength of the $C-H$ bond that is broken, reflecting the enthalpic factor, and this is the reason that the relative reactivity of H-atom abstraction increases in the order: primary < secondary < tertiary, as indicated by the data in Table 10.3. (The bond strength order D_{C-H} decreases in the order $H-CH_2R > H-CHR_2 > H-CR_3$, reflecting the greater stability of the more substituted alkyl radicals). Note, however, that the selectivity of the radical increases dramatically as the D_{X-H} bond strength decreases. Thus in the highly exothermic reaction with $F\cdot$, selectivity is low, while for the endothermic reaction with $I\cdot$, selectivity is high. This reaction series provides a particularly clear illustration of the *reactivity–selectivity principle* (see Section 5.9).

The importance of polar effects in radical abstraction reactions, arising from the mixing of charge-transfer configurations, have also been demonstrated. For example, the rate of abstraction of a benzylic hydrogen atom from a family of substituted toluenes

$$Y^{\bullet} \ + \quad \text{(substituted toluene, } CH_3, X) \quad \longrightarrow \quad Y{-}H \ + \quad \text{(radical, } \overset{\bullet}{C}H_2, X) \qquad (10.13)$$

TABLE 10.3. Relative Reactivity of Primary, Secondary, and Tertiary Hydrogens in Radical Abstraction Reactions[a]

Radical	Temperature (°C)	$H-CH_2R$	$H-CHR_2$	$H-CR_3$	D_{X-H}
$F\cdot$	25	1	1	2	136
$Cl\cdot$	25	1	4	7	103
$Br\cdot$	150	1	80	2,000	87
$I\cdot$	150	1	1,000	97,000	71

[a]Data taken from J. M. Tedder, *Angew. Chem. Int. Ed. Engl.* **21**, 401 (1982).

for $Y\cdot = Cl\cdot$, $Br\cdot$, $Cl_3C\cdot$, and $C_6H_5^{\cdot}$ each follows a Hammett correlation with a negative ρ value. This suggests a polar contribution to the transition state, as described by

$$\overset{\delta-}{Y}\text{---}H\text{---}\overset{\delta+}{CH_2}$$

10.15

or in configuration-mixing terms terms by

$$\Psi_{TS} = DA \leftrightarrow D^3A* \leftrightarrow D^-A^+ \tag{10.14}$$

The possibility that in this case the ring substituent effect is, at least partly enthalpic, cannot be ruled out since it turns out that electron-releasing substituents in the ring not only *increase* charge transfer to the attacking radical, but also *weaken* the C—H bond. In other words, the rate-enhancing effect of electron-releasing substituents may be partly enthalpic in nature, and not just due to the polar effect. In other cases, however, the role of polar effects is more clear-cut. The rate of hydrogen abstraction from alkanes by $Cl\cdot$ is greater than by $\cdot CH_3$ despite the fact that the reaction enthalpies are essentially the same; H—Cl and H—CH$_3$ bond strengths are almost the same (103 and 104 kcal mol^{-1}, respectively). So in this case polar effects are expressed explicitly.

In summary, radical reaction energy profiles, like other reaction types, may also be built up from the mixing of VB configurations. In contrast to the polar reactions (Chapter 9) where the charge-transfer configuration was the *product* configuration, for many radical reactions it constitutes an "intermediate" configuration. This explains the large polar character that often characterizes the transition states of radical processes, but which does not show up in the reaction products themselves.

10.4 PERICYCLIC REACTIVITY

Pericyclic reactions are a class of concerted reactions in which the bonding changes that take place occur within cyclic transition states. The factors that govern pericyclic reactivity have been of intense theoretical interest over the past 30 years, following Woodward and Hoffmann's[2] landmark contribution that orbital symmetry, till then an obscure theoretical abstraction, determined whether a particular reaction was "allowed" or "forbidden."

Extensive coverage of the mechanism of pericyclic reactions, and their subdivision into cycloaddition, sigmatropic, electrocyclic, cheletropic reactions, and group transfers, can be found in many current texts. These invariably discuss the reactions in MO terms. In this chapter we discuss the class of cycloaddition reactions in both MO and VB terms, and show the way in which the two approaches relate to one another. In particular the reason cycloaddition reactions have an activation barrier, *even those that are symmetry allowed*, will be explored.

10.4.1 *4n + 2* Rule

Consider the dimerization of two ethylenes to give cyclobutane

$$(2 + 2) \quad \| \; + \; \| \longrightarrow \square \qquad (10.15)$$

and the cycloaddition of ethylene and butadiene to give cyclohexene

$$(4 + 2) \qquad \longrightarrow \qquad (10.16)$$

Both of these reactions involve an electronic reorganization in which four π electrons reorganize into two new σ bonds. Both can be readily described by the curved arrow representation, shown in Eq. (10.16) for the cycloaddition reaction. However, the curved arrow representation is of limited mechanistic value since it fails to explain the large reactivity differences between the two systems; the barrier to the ethylene dimerization reaction [Eq. (10.15)] is particularly high despite the fact that the reaction is strongly favored enthalpically ($\Delta H° = -18.2$ kcal mol^{-1}), while the cycloaddition reaction of ethylene and butadiene [Eq. (10.16)] and its derivatives (the Diels–Alder reaction) is relatively facile and represents one of the more important synthetic reactions in organic chemistry.

Woodward and Hoffmann pointed out that two factors govern cycloaddition reactivity: (a) the *number* of π electrons in the reacting system and (b) the stereochemical approach of the two substrates toward one another. Let us discuss these factors in turn.

The number of π electrons is central in governing cycloaddition reactivity. If the system contains a total of *4n* π electrons (where *n* is an integer, 1, 2, 3, . . .) the reaction is "forbidden." However, if the system contains a total of *4n + 2* π electrons, the reaction is "allowed." For this reason the dimerization of two ethylenes with 4π electrons (i.e., a *4n* system; $n = 1$) is "forbidden," while the cycloaddition of ethylene and butadiene, with 6π electrons (i.e., a *4n + 2* system; $n = 1$) is "allowed." Because of the importance of the electron count, cycloaddition reactions are classified by the number of π electrons in each of the two reacting components. Thus the ethylene dimerization reaction

[Eq. (10.15)] is termed a (2 + 2) cycloaddition, while the reaction of ethylene with butadiene [Eq. (10.16)] is termed a (4 + 2) cycloaddition.

Why is the number of π electrons so important in determining reactivity? It turns out that the symmetry properties of the molecules change with the number of π electrons and these have a direct bearing on the energy of the transition state. There are different ways of demonstrating this symmetry dependence, the simplest being based on frontier MO theory.

Consider the face-on (termed *suprafacial*) approach of two ethylene molecules, illustrated in **10.16,** which allows the two π systems to interact with one another. An orbital interaction diagram that describes the possible interactions that may occur between the frontier orbitals, is shown in Fig. 10.5. At first sight three interactions appear to be possible: a four-electron π–π interaction (HOMO–HOMO) (labeled *a*), which is destabilizing, and two two-electron π–π^* (HOMO–LUMO) interactions, which potentially would be stabilizing (labeled *b* and *c*). However, as described in Section 2.2.3, for the suprafacial approach of two π-systems, π–π^* overlap is zero due to a symmetry mismatch. As a result, the only interaction that takes place between the two π systems is the *destabilizing* one: interaction *a*. We conclude therefore that the suprafacial approach of two ethylene molecules, as indicated in **10.16,** will be repulsive.

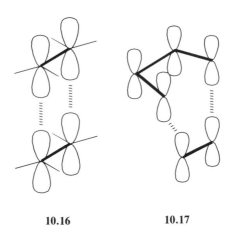

10.16   10.17

Now consider the suprafacial approach of ethylene and butadiene as illustrated in **10.17.** Here predominant interaction is between the butadiene π system at C1 and C4 and the ethylene π system. The corresponding orbital interaction diagram for this reaction is illustrated in Fig. 10.6. In this case both of the two-electron *stabilizing* interactions (HOMO–LUMO) *b* and *c* are now *symmetry allowed* while the four-electron *destabilizing* interaction (HOMO–HOMO) *a* is *symmetry forbidden. Thus increasing the number of electrons in the π system by two has changed the orbital topology so that symmetry-allowed interactions become symmetry forbidden and symmetry-forbidden interactions become symmetry allowed.* The result: changing from the two ethylene system

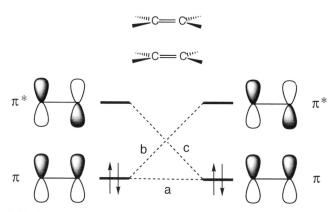

Fig. 10.5. Orbital interaction diagram illustrating the repulsive interaction that arises from the face-on approach of two ethylene molecules. Potentially stabilizing two-electron interactions *b* and *c* are symmetry forbidden, while the destabilizing four-electron interaction *a* is symmetry allowed.

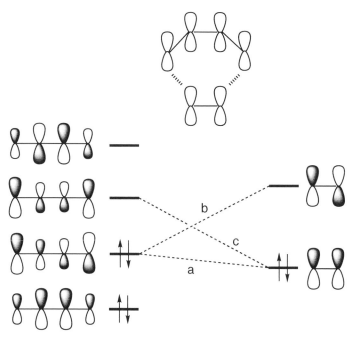

Fig. 10.6. Orbital interaction diagram illustrating the stabilizing interactions that arises from the face-on approach of ethylene and butadiene. Stabilizing two-electron interactions *b* and *c* are symmetry allowed while the potentially destabilizing four-electron interaction *a* is symmetry forbidden.

[Eq. (10.15)] to the butadiene and ethylene system [Eq. (10.16)] has resulted in the HOMO–HOMO *destabilizing* interaction being turned off, while the two HOMO–LUMO *stabilizing* interactions are turned on.

The role of electron count may be described as follows: *If the interacting system involves an odd number of electron pairs (4n + 2 electrons), then suprafacial stabilizing HOMO–LUMO orbital interactions are symmetry allowed and suprafacial destabilizing HOMO–HOMO interactions are symmetry forbidden. If an even number of electron pairs are involved (4n electrons), then destablizing HOMO–HOMO orbital interactions are symmetry allowed and stabilizing HOMO–LUMO interactions are symmetry forbidden.*

The above statement can be formulated in reactivity terms as follows: *for suprafacial attack, pericyclic reaction with 4n + 2 electrons are thermally allowed, while those with 4n electrons are thermally forbidden.* Interestingly, we can now see that symmetry-allowed cycloaddition and aromaticity (two apparently unrelated concepts) are actually intimately related. Both phenomena may be understood in terms of electron count and its symmetry consequences; *it is the same electron count factor that explains the stability of benzene compared to cyclobutadiene and the relatively facile cycloaddition of ethylene and butadiene compared to the high barrier to ethylene dimerization.* Thus systems with 2, 6, 10, 14, . . . π electrons will show aromatic stabilization and this may take place in the ground state (as in aromatic compounds) or in the transition state (as in cycloaddition). Conversely, systems with 4, 8, 12, 16, . . . π electrons will exhibit antiaromatic destablization, in both the ground and transition states.

We noted earlier that reaction stereochemistry, and not just electron count, is important in governing pericyclic reactivity. Let us now elaborate on this point. The presence of a π system within a molecule causes it to possess two faces. Reaction of the π system may, in principle, take place at one face (in a *suprafacial* mode), as shown by the arrows in **10.18,** or at both faces (in an *antarafacial* mode), as shown by the arrows in **10.19.** Because of the antisymmetric character of all π systems, *suprafacial and antarafacial approaches have opposite symmetry requirements. If suprafacial attack is symmetry allowed, then antarafacial attack is symmetry forbidden and vice versa.*

For cycloaddition reactions, where two molecules are involved, each mol-

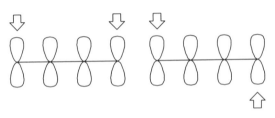

Suprafacial attack   Antarafacial attack

10.18   **10.19**

ecule may, in principle, react in a suprafacial or antarafacial mode. Thus the *4n + 2* rule, as stated above, applies to the most common situation, where both molecules react in a suprafacial mode. However, if one of the molecules reacts in an antarafacial mode, then the rule must be reversed; reactions of systems with *4n + 2* electrons will be forbidden and those with *4n* electrons will be allowed. For the unusual case where *both* molecules react antarafacially, then *4n + 2* systems will be allowed, just as for suprafacial–suprafacial attack.

Examples of cycloaddition reactions that illustrate the *4n + 2* rule are shown in Eqs. (10.17–10.20). Note that the reaction of heptafulvalene and tetracyanoethylene, Eq. (10.21), a (14 + 2) system belongs to the *4n* class and might therefore be considered to be forbidden. In fact, from the trans stereochemistry of the product it is clear that attack at the heptafulvalene has occurred *antarafacially*, rather than suprafacially, so the (14 + 2) cycloaddition reaction, predicted to be forbidden in suprafacial–suprafacial mode, is actually antarafacial–suprafacial, and hence symmetry allowed.

While the orbital interaction diagram is particularly effective in explaining why some cycloaddition reactions proceed readily while others do not, the question arises *Why do allowed cycloaddition reactions have a barrier at all?* After all, if these reactions are governed by stabilizing two-electron interactions, then what is the source of the reaction barrier? A curve-crossing analysis of the reaction provides insight into this question and provides a somewhat different perspective on the role of symmetry in pericyclic reactions.

$$(4 + 2) \qquad\qquad\qquad\qquad\qquad\qquad (10.17)$$

$$(4 + 6) \qquad\qquad\qquad\qquad\qquad\qquad (10.18)$$

$$(4 + 2) \qquad\qquad\qquad\qquad\qquad\qquad (10.19)$$

$(4 + 2)$  $\qquad$ (10.20)

$(14 + 2)$ 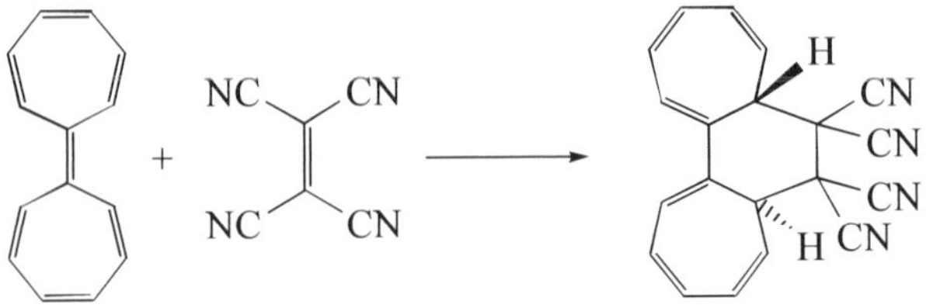 $\qquad$ (10.21)

10.4.2   A Curve-Crossing Approach to Cycloaddition Reactivity

Let us specify the key electronic configurations that control the dimerization reaction of ethylene so that we can build up an energy profile for this reaction and obtain additional insight into the factors that govern reactivity in generalized cycloaddition systems.[3] Electronically, this reaction is very similar to the bond exchange process

$$H_2 + D_2 \rightarrow 2HD \qquad (10.22)$$

described in Section 4.4.3. Both the bond-exchange process and cycloaddition involve the breaking of two existing bonds and the making of two new bonds.

The key VB configurations for the ethylene cycloaddition are illustrated in **10.20–10.22**, while the key MO configurations are illustrated in **10.23–10.25**. The VB configuration **10.20**, labeled DA, is termed the reactant configuration because this configuration is a good descriptor of the reactants; the two electrons on each ethylene moiety are spin-paired to form the two π bonds while the two electrons on *different* ethylene moieties have the same spin and so are *not* coupled to one another. Configuration **10.23** is just the MO representation of this VB configuration; the π orbital of each ethylene is occupied by an electron pair.

Configuration **10.21** is the VB product configuration. As for **10.20**, each carbon is still associated with one electron, however, the spin arrangement is different; the electron pairs are now coupled up to form the two cyclobutane σ bonds, and *not* the ethylene π bonds. In order to obtain this configuration from the reactant configuration **10.20**, each of the π-bond electron pairs needs to be uncoupled. The spectroscopic state that most closely describes this uncoupled situation is the triplet state. *Thus each of the π bonds needs to be excited from the singlet to the triplet state.* Accordingly, this configuration is labeled $^3D^* {}^3A^*$. An MO representation of this VB configuration is shown in **10.24**.

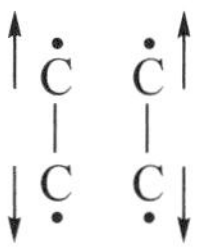

DA

$^3D^* \, ^3A^*$

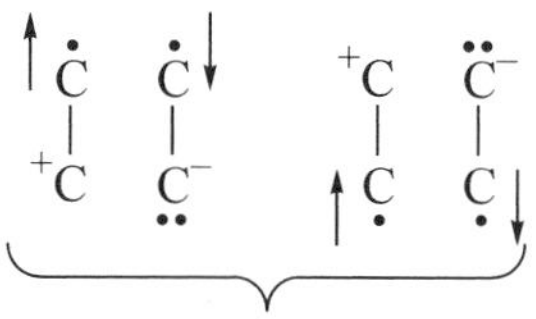

D^+A^-

10.20   **10.21**   **10.22**

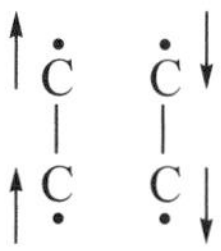

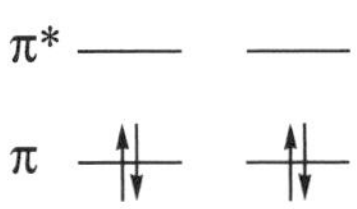

π^* —— —— π^* ↑ ↓ π^* —— ↓

π ↑↓ ↑↓ π ↑ ↓ π ↑ ↑↓

DA $^3D^* \, ^3A^*$ D^+A^-

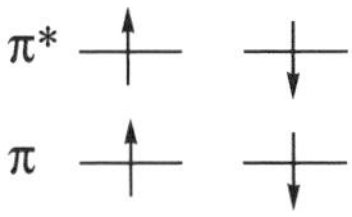

10.23   **10.24**   **10.25**

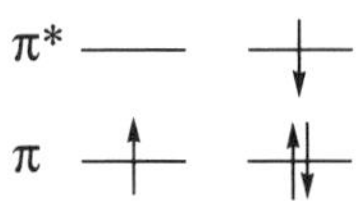

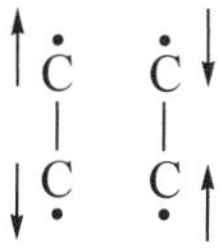

10.26

Note that $^3D^* \, ^3A^*$ is an overall singlet configuration despite the fact it contains within it two local triplets.[a]

Finally, one might consider the possible involvement of charge-transfer configurations as well. The basic one, labeled D^+A^-, is shown as a VB configuration in **10.22** and as an MO configuration in **10.25**. Of course, in addition to D^+A^- there is also the equivalent configuration D^-A^+, which describes an electron transfer from A to D, rather than from D to A (clearly, for two ethylene molecules the labels of D and A are arbitrary so that D^+A^- and D^-A^+, being equal in energy, would contribute equally to the ground-state profile).

An energy profile for ethylene dimerization, built up schematically from the basis set of VB configurations, is illustrated in Fig. 10.7. The reactant configuration DA rises in energy as a function of the reaction coordinate because approach of the two ethylene moieties induces triplet repulsion between the odd electrons on the two ethylenes that are not coupled to one another. In

---

[a]At this point it might naively seem that there is a third way of setting the spins of four electrons in a square arrangement, shown in **10.26**. This configuration is also a singlet configuration, however, each electron appears to be spin paired to *two* neighbors. In fact this representation does not describe a third state; there are only *two* quantum mechanically allowed singlet states for a four-electron system, and these are best represented by **10.20** and **10.21**. Configuration **10.26** can be thought of as a linear combination of these two proper spin representations.[4]

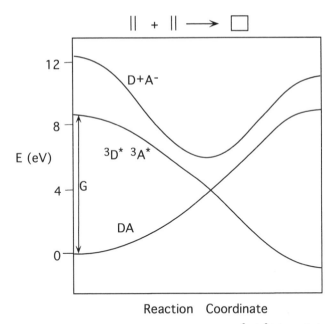

Fig. 10.7. Configuration curves for reactant DA, product $^3D^* \, ^3A^*$, and charge-transfer D^+A^-, configurations for the dimerization reaction of ethylene to give cyclobutane. Mixing of D^+A^- into the ground-state wave function is precluded by symmetry so that a large barrier for this reaction would result (see text). $\|$, ethylene, $\square$, cyclobutane.

contrast, $^3D^* \, ^3A^*$ starts off high in energy—at approximately 8.6 eV on the energy scale (the vertical singlet–triplet excitation energy for ethylene is 4.3 eV and since both ethylene molecules need to be excited, this value is doubled). However, as we move along the reaction coordinate the energy drops sharply since the uncoupled electrons in the two ethylene moieties can couple together to form the two new σ bonds of the cyclobutane product in an exothermic reaction ($\Delta H° = -18.2$ kcal mol^{-1}). *It is the avoided crossing of these two configurations that leads to the simplest description of the ground (and excited)-state energy profiles for ethylene dimerization.*

In order to improve our description of the ground-state energy profile we should now consider the incorporation of an "intermediate" configuration into our description of the reaction profile. The obvious candidate is the charge-transfer configuration D^+A^- (and, of course, the isoenergetic D^-A^+). The energy of D^+A^- as a function of the reaction coordinate is also shown in Fig. 10.7. Initially it is high in energy—the ionization energy and electron affinity of ethylene are 10.5 and -1.8 eV, respectively, placing D^+A^- initially at about 12.3 eV. However, its energy drops sharply as the two ethylenes approach one another [by ~ 6 eV when the two charges are 2.4 Å apart; see Eq. (4.3) in Chapter 4] so that the charge-transfer configuration comes quite close in energy to the crossing point of DA and $^3D^* \, ^3A^*$. It therefore may, in principle, contribute to a description of the reaction profile. *However, for the cycloaddition of two ethylene moieties the symmetry of D^+A^- and DA precludes mixing.* DA and D^+A^- cannot mix because their mixing would be governed by the mixing of π and π^* orbitals (see Section 2.2.3) and for two molecules of ethylene these two orbitals are orthogonal. The above statements may be expressed mathematically (see Section 3.4) as

$$\langle DA \,|\mathcal{3C}|\, D^+A^- \rangle \approx \langle \pi |\mathcal{3C}| \pi^* \rangle \approx 0 \qquad (10.23)$$

In view of the symmetry mismatch between DA and D^+A^- that precludes mixing of these two configurations, a concerted cycloaddition of two molecules of ethylene, if it were to occur, would involve a DA $-$ $^3D^*$ $^3A^*$ avoided crossing. Given the large initial energy gap between DA and $^3D^*$ $^3A^*$ (~ 200 kcal mol^{-1}) the resultant barrier would be very high. Even assuming a low f factor of 0.2 in the curve-crossing equation implies a barrier of about 40 kcal mol^{-1}. (For the rate implications of different barrier heights see Table 5.1.) Thus the fact that ethylene cycloaddition does not occur thermally in a concerted process can be understood.

Let us now turn to the cycloaddition of ethylene and butadiene. For this system the curve-crossing diagram would be similar in appearance to that shown in Fig. 10.7, though both D^+A^- and $^3D^*$ $^3A^*$ are lower in energy than for the ethylene dimerization reaction; D^+A^- is lower because the ionization potential of butadiene (9.1 eV) is lower than that for ethylene (10.5 eV), while $^3D^*$ $^3A^*$ is lower because the singlet–triplet excitation energy for a conjugated system (butadiene) is less than that for ethylene (in butadiene the HOMO level is higher and the LUMO level is lower than in ethylene). *Most importantly, however, the symmetry of D^+A^- now enables it to mix into the ground-state wave function;* DA–D^+A^- mixing is allowed because for the ethylene–butadiene system, HOMO–LUMO overlap *is* symmetry allowed. Thus the wave function for the transition state for butadiene–ethylene cycloaddition allows for charge-transfer mixing and may be described by

$$TS = c_1(DA) + c_2(D^+A^-) + c_3(^3D^*\ ^3A^*) \qquad (10.24)$$

Mixing in of the charge-transfer configuration both *stabilizes* the transition state and provides it with charge-transfer character. We see therefore that the role of symmetry in governing the rate of cycloaddition reactions manifests itself in essentially the same way through the curve-crossing model as it does through an FMO analysis.

Certain experimental features associated with cycloaddition reactions can be understood in these terms. It is well known that electron-releasing substituents in the diene and electron-withdrawing substituents in the dienophile enhance cycloaddition reactivity. Thus methyl-substituted butadienes and cyanoethylene are more reactive than butadiene and ethylene. In MO terms the enhanced reactivity is explained by the effect of the substituents on the HOMO–LUMO energy gap. A methyl substituent raises the energy of the diene HOMO while the cyano group lowers the energy of the ene LUMO: the net result is a reduced HOMO–LUMO gap and an enhanced two-electron stabilization energy.

Substituent effects on cycloaddition reactivity can be understood in configuration terms as well. In configuration terms we would say that reactivity is dominated by the energy of the charge-transfer configuration D^+A^-. Substituents that stabilize D^+A^- will cause the transition state to take on more charge-transfer character leading to enhanced stabilization. Since the energy of D^+A^- is governed by the ionization potential of D (the diene) and the electron affinity of A (the dienophile), substituents in the diene that reduce the diene ionization potential and substituents in the dieneophile that increase the dienophile electron affinity will lower the energy of D^+A^- and thereby enhance reactivity. This is illustrated in Fig. 10.8, which shows DA, $^3D^* \, ^3A^*$, and D^+A^- configurations (labeled in the figure) together with the ground-state reaction profile (firm bold line). The effect of a stabilizing substituent on the energy of D^+A^- is indicated by the light broken line. The result of the enhanced mixing of this lower energy D^+A^- configuration leads to the stabilization of the entire ground-state reaction profile, as shown by the bold broken line. Thus, in configuration terms we would say that methylbutadiene and cyanoethylene are more reactive than butadiene and ethylene since these substituents lead to a lowering of the energy of D^+A^-, and therefore to more charge-transfer mixing in the transition state.

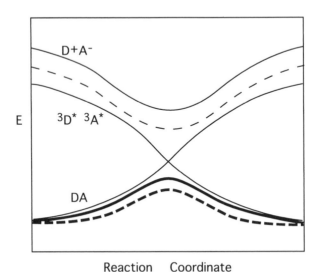

Reaction   Coordinate

FIg. 10.8. Curve-crossing diagram for an allowed cycloaddition reaction showing the effect of stabilizing D^+A^- on the energy of the ground-state reaction profile. The stabilized D^+A^- curve is shown by the light dashed line; the ground-state profile for the unsubstituted reaction is shown by the bold line, and with the stabilizing substituent, by the dashed bold line.

In summary the curve-crossing model predicts that the factors that will govern reactivity in cycloaddition reactions are

1. The singlet–triplet excitation energies for each of the reactant alkenes (these values will determine the initial energy gap between reactant and product configurations G in Fig. 10.7, and hence the energy of the top-left-hand anchor point in the curve-crossing diagram).

2. The thermodynamic stability of the product (this determines the energy of the bottom-right-hand anchor point in the curve-crossing diagram).

3. The energy of charge-transfer configurations (D^+A^- and D^-A^+) in the transition state region. This parameter appears to be a dominant factor in governing reactivity in symmetry-allowed cycloadditions.

4. The symmetry properties of the charge-transfer configurations, which will determine whether these configurations can mix into the ground-state reaction profile. *It is the symmetry properties of the charge-transfer configurations that form the basis for the Woodward–Hoffmann rules.*

An interesting prediction of the curve-crossing analysis is the possibility of intermediate formation in allowed cycloaddition reactions. Recall that a single-step process may become a two-step process, when there is an intermediate configuration of sufficiently low energy that it dominates the ground-state reaction profile at some intermediate geometry (see Fig. 4.11, Section 4.4.4). *For symmetry-allowed cycloaddition reactions this means that if D^+A^- is strongly stabilized then a cycloaddition reaction may proceed in a step-wise fashion.* In curve-crossing terms the first step of the reaction would be described by the avoided crossing of DA and D^+A^-, while the second step would result from the avoided crossing of D^+A^- and $^3D^*$ $^3A^*$. In chemical terms this means the reaction *would involve an initial electron transfer to generate a radical cation and a radical anion.* These two species would then combine to form the cycloadduct.

An example of a cycloaddition reaction that proceeds in this fashion was recently reported and is shown in Eq. (10.25).[5] In this case, we have a particularly good donor–acceptor pair—the dialkoxytrimethylenemethane (**10.27**) has a very low ionization potential and the nitroalkene (**10.28**) has a high electron affinity so the energy of D^+A^- is particularly low. Indeed the electron-transfer route is only observed with alkenes whose reduction potentials are larger than -1.8 V. When **10.27** reacts with *less* electrophilic alkenes (i.e., so that the energy of the D^+A^- configuration would be higher), a conventional concerted cycloaddition mechanism is observed. Therefore we see that the curve-crossing treatment of cycloaddition can explain the change in mechanism of an *allowed* cycloaddition reaction from a *concerted* pathway to a *two-step* pathway via electron-transfer intermediates.

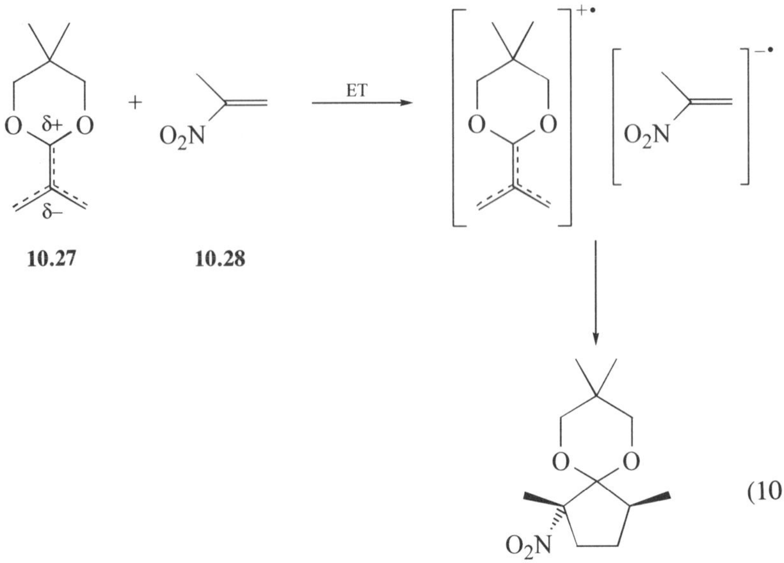

(10.25)

10.29

10.5   A CURVE-CROSSING APPROACH TO CARBENE REACTIVITY

A configuration analysis of cycloaddition reactions and carbene reactions reveals that from a theoretical point of view these two reaction classes are very similar.[6] Consider the addition of a singlet carbene to an alkene to generate a cyclopropane

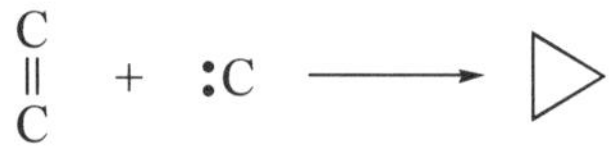

$$
\begin{matrix} C \\ \| \\ C \end{matrix} \quad + \quad :C \quad \longrightarrow \quad \triangle \qquad (10.26)
$$

The important VB configurations that contribute to the reaction profile wave function for this reaction are shown in **10.30–10.33**. The reactant configuration DA (**10.30**) describes the electron distribution in the reactants—a π-electron pair in the alkene, and a lone-pair orbital in the singlet carbene—while the product configuration $^3D*\,^3A*$ (**10.31**) shows the electron distribution in the products; the electronic reorganization involves spin changes that allow electron coupling to generate the two new σ bonds (just as for cycloaddition). Thus $^3D*\,^3A*$ may be obtained from DA by the excitation of both the alkene and the singlet carbene to their corresponding triplet states—hence the label $^3D*\,^3A*$. Charge-transfer configurations D^+A^- and D^-A^+, illustrated in **10.32** and **10.33**, are intermediate configurations and may mix into the ground-state wave function when they are sufficiently low in energy.

D A	^{3}D* ^{3}A*	D$^+$ A$^-$	D$^-$ A$^+$
10.30	**10.31**	**10.32**	**10.33**

The analogy with cycloaddition reactions, which also rely on the set of DA, ^{3}D* ^{3}A*, D$^+$A$^-$, and D$^-$A$^+$ configurations **10.20–10.22**, is apparent. Indeed, by analogy with cycloaddition reactions, carbene reactivity is often dominated by the energy of the two charge-transfer configurations. Thus for most carbenes, the energy of D$^+$A$^-$ governs reactivity, that is, carbenes tend to be *electrophilic* (e.g., CCl$_2$, CBr$_2$, CH$_3$CCl, and PhCCl). However, for carbenes with low ionization potentials [e.g., (CH$_3$O)$_2$C], it is D$^-$A$^+$ that dominates reactivity so that these carbenes display *nucleophilic* character. When the energies of D$^+$A$^-$ and D$^-$A$^+$ are relatively similar, as for CH$_3$OCCl and PhOCCl, the carbene will display *ambiphilic* character. Thus ambiphilic carbenes react rapidly with both nucleophilic and electrophilic alkenes, since both D$^+$A$^-$ and D$^-$A$^+$ are low in energy and may contribute to a description of the transition state wave function. Note, however, that since the energy of D$^+$A$^-$ and D$^-$A$^+$ depend not only on the nature of the carbene, but on the electronic character of the alkene that is attacked (its I and A values), classification of a carbene as electrophilic or nucleophilic is not absolute; the electrophilicity or nucleophilicity of a particular carbene will also be influenced by the character of the substrate alkene with which it reacts.

Finally, as might be expected from the above description, carbene reactivity appears to depend not just on charge-transfer energies, D$^+$A$^-$ and D$^-$A$^+$, but on the energy of the ^{3}D* ^{3}A* configuration as well. The configuration-mixing diagram that was built up for cycloaddition (Fig. 10.7), may also be used to describe a carbene addition reaction. The energy of ^{3}D* ^{3}A* is determined by (a) the initial energy gap (top-left-hand anchor point), which is governed by the carbene and alkene singlet–triplet energy gaps and (b) by the reaction exothermicity (bottom-right-hand anchor point). Indeed CH$_2$, whose singlet–triplet gap is actually negative ($\sim$ −9 kcal mol^{-1}), and whose reaction with alkenes is highly exothermic (−105 kcal mol^{-1}) is thought to have a *negative* energy of activation, whereas CF$_2$, whose singlet–triplet gap is large (57 kcal mol^{-1}) and whose heat of reaction is less negative (−47 kcal mol^{-1}) is found to have a relatively large activation barrier of 24–27 kcal mol^{-1}.

In conclusion we see that there is a strong theoretical similarity between cycloaddition and carbene addition reactions. Accordingly, a configuration mixing diagram for cycloaddition, such as that illustrated in Fig. 10.7, can also be used to qualitatively predict substituent effects on reactivity in carbene reactions.

REFERENCES

1. For applications of the curve-crossing model to radical additions, see (a) S. S. Shaik and P. C. Hiberty, *J. Am. Chem. Soc.* **107,** 3089 (1985); (b) A. Pross, *Isr. J. Chem.* **26,** 390 (1985); (c) S. S. Shaik, P. C. Hiberty, J.-M. Lefour, and G. Ohanessian, *J. Am. Chem. Soc.* **109,** 363 (1987); (d) S. S. Shaik and E. Canadell, *J. Am. Chem. Soc.* **112,** 1446 (1990); (e) M. W. Wong, A. Pross, and L. Radom, *J. Am. Chem. Soc.* **115,** 11050 (1993); (f) M. W. Wong, A. Pross, and L. Radom, *J. Am. Chem. Soc.* **116,** 6284 (1994); **116,** 11938 (1994).

2. R. B. Woodward and R. Hoffmann, *The Conservation of Orbital Symmetry*, Academic, New York, 1970.

3. (a) N. D. Epiotis, S. S. Shaik, and W. Zander. In *Rearrangements in Ground and Excited States*, P. De Mayo, Ed., Vol. 2. Academic, New York, 1980; (b) A. Pross, *Adv. Phys. Org. Chem.* **21,** 99 (1985); (c) A. Ioffe and S. Shaik, *J. Chem. Soc. Perkin Trans. 2*, 2101 (1992).

4. For a summary of allowed spin states, see L. Salem, *Electrons in Chemical Reactions: First Principles*, Wiley, New York, 1982, p. 8–9.

5. S. Yamago, S. Ejiri, M. Nakamura, and E. Nakamura, *J. Am. Chem. Soc.* **115,** 5344 (1993).

6. A. Pross and R. A. Moss, *Tetrahedron Letters*, 4553 (1990). For a recent review of carbene addition to alkenes see R. A. Moss, *Acc. Chem. Res.* **22,** 15 (1989).

INDEX

Of